First Book of
Indian Botany

DANIEL OLIVER

CAMBRIDGE
UNIVERSITY PRESS

CAMBRIDGE UNIVERSITY PRESS

Cambridge, New York, Melbourne, Madrid, Cape Town,
Singapore, São Paolo, Delhi, Mexico City

Published in the United States of America by Cambridge University Press, New York

www.cambridge.org
Information on this title: www.cambridge.org/9781108055628

© in this compilation Cambridge University Press 2013

This edition first published 1869
This digitally printed version 2013

ISBN 978-1-108-05562-8 Paperback

CAMBRIDGE LIBRARY COLLECTION

Books of enduring scholarly value

Life Sciences

Until the nineteenth century, the various subjects now known as the life sciences were regarded either as arcane studies which had little impact on ordinary daily life, or as a genteel hobby for the leisured classes. The increasing academic rigour and systematisation brought to the study of botany, zoology and other disciplines, and their adoption in university curricula, are reflected in the books reissued in this series.

First Book of Indian Botany

Well known among his contemporaries for his unrivalled knowledge of aberrant plants, Daniel Oliver (1830–1916) ran the herbarium at Kew Gardens and held the chair of botany at University College London, for which he was recommended by Charles Darwin. Although Oliver never visited India, his expertise in Indian botany grew considerably after he worked with an enormous number of dried specimens rescued from the cellars of the East India Company. In this book, first published in 1869, he sets out the basics of botanical study in India for the absolute beginner. It includes instruction on the anatomy of simple plants, lessons in collection and dissection, and explanations of botany's often dense terminology. Annotated diagrams appear throughout, in both microscopic and macroscopic views. Rigorous and carefully structured, Oliver's book remains an excellent resource for novice botanists and students in the history of science.

Cambridge University Press has long been a pioneer in the reissuing of out-of-print titles from its own backlist, producing digital reprints of books that are still sought after by scholars and students but could not be reprinted economically using traditional technology. The Cambridge Library Collection extends this activity to a wider range of books which are still of importance to researchers and professionals, either for the source material they contain, or as landmarks in the history of their academic discipline.

Drawing from the world-renowned collections in the Cambridge University Library and other partner libraries, and guided by the advice of experts in each subject area, Cambridge University Press is using state-of-the-art scanning machines in its own Printing House to capture the content of each book selected for inclusion. The files are processed to give a consistently clear, crisp image, and the books finished to the high quality standard for which the Press is recognised around the world. The latest print-on-demand technology ensures that the books will remain available indefinitely, and that orders for single or multiple copies can quickly be supplied.

The Cambridge Library Collection brings back to life books of enduring scholarly value (including out-of-copyright works originally issued by other publishers) across a wide range of disciplines in the humanities and social sciences and in science and technology.

FIRST BOOK

OF

INDIAN BOTANY.

FIRST BOOK

OF

INDIAN BOTANY.

BY

DANIEL OLIVER, F.R.S., F.L.S.

KEEPER OF THE HERBARIUM AND LIBRARY OF THE ROYAL GARDENS, KEW, AND
PROFESSOR OF BOTANY IN UNIVERSITY COLLEGE, LONDON.

WITH NUMEROUS ILLUSTRATIONS.

London:
MACMILLAN & CO.
1869.

LONDON :

R. CLAY, SONS, AND TAYLOR, PRINTERS,

BREAD STREET HILL.

PREFACE.

THIS little book is, in substance, my "Lessons in Elementary Botany" adapted for use in India.

But in preparing it, I have had in view the want, often felt, of some handy *résumé* of Indian Botany, which might be serviceable not only to residents in India, but also to any one about to proceed thither, desirous of getting some preliminary idea of the Botany of that country.

I might have entitled the book "Illustrations of Indian Natural Orders of Plants;" but as the same chapters, with necessary alteration, on the Elements of Structural and Physiological Botany are prefixed to the systematic part which I originally drew up for my previous work, the whole thing seems to me tolerably suited to serve as a "First Book of Indian Botany."

My chief difficulty has been in the selection of suitable Types to illustrate the Natural Orders on the plan of the late Professor Henslow. In a book for use in Britain there is no difficulty on this head, for so many plants are pretty uniformly dispersed over our limited area, that nearly every

Type-species is within reach throughout our islands. But in India the case is very different. Few easily recognisable species have a sufficiently general distribution, or, if they have, it often happens that some exceptional peculiarity of structure unfits them for service as typical examples of any group. One consequence of this difficulty is, that I have been obliged to use, in some cases, garden plants which are widely cultivated in India, although not actually indigenous there, in preference to native species. For example, in the large Order Compositæ I have preferred the Mexican *Zinnia*, so generally grown in Indian gardens, to any native species ; and the Chinese *Dendrobium nobile* to any Indian species of the great Orchid Family. This latter is perhaps unfortunate. There are a few wild Indian Orchids with a wide distribution, but they are either difficult of recognition by those not accustomed to botanical pursuits, or unfitted in some other way to serve my purpose. However, in such cases as these I have nearly always introduced cuts, drawn by Mr. Fitch, which I trust may partially substitute the actual specimens, if these be not obtainable.

With regard to the selection of suitable and easily accessible illustrative Types, I shall be glad to receive advice from any one interested in the matter. Perhaps it would be a good plan to introduce two, three, or more species in illustration of the different Orders in cases in which the common conspicuous species are restricted in their geographical range. As to this, I may find reason to modify the book, should it ever get, as I hope it may, into a second edition.

I have never visited India myself. I have, however, under Dr. Hooker's direction, had much to do with Indian Botany, especially in connexion with the enormous store (about twelve or fourteen large waggon-loads) of dried Indian plants rescued from the cellars of the late East India Company, through the persevering efforts of this distinguished Botanist, which have been arranged and distributed at Kew during the last ten years.

In writing out this little work, I have been greatly indebted to Dr. Hooker, and to Dr. T. Thomson, of Kew, for many useful hints, especially concerning the selection of illustrative Type-species. Dr. Thomson and Mr. Edgeworth have also very kindly given me much help in introducing some local or native names of commoner species; a matter upon which I have no personal knowledge.

KEW, LONDON, *August* 1869.

CONTENTS.

PART I.

THE ELEMENTS OF STRUCTURAL AND PHYSIOLOGICAL BOTANY.

CHAPTER VI.

CHAPTER VII.

CHAPTER VIII.

PART II.

FIRST BOOK OF INDIAN BOTANY.

CHAPTER I.

CHAPTER II.

CHAPTER III.

CHAPTER IV.

CHAPTER V

APPENDIX.

PART I.

THE ELEMENTS OF STRUCTURAL AND PHYSIOLOGICAL BOTANY.

CHAPTER I.

OF THE ROOT, STEM, AND LEAVES.

1. The Root.—Its form and general structure : it penetrates the soil ; is colourless ; irregularly branched ; destitute of leaves ; and its extremities are sheathed.
2. The Stem.—It ascends ; is coloured ; bears leaves and branches at definite points ; the extremities are not sheathed, but give off, successively, rudiments of leaves.
3. Foliage-leaves.—They are borne by the stem only. " Radical " and " cauline " leaves : thin, coloured green, consisting of a horizontally expanded blade with, or without, a petiole.
4. Flowers consist of leaves. The peduncle. Suppression of internodes in flowers. The receptacle of the flower.
5. The Sepals ; forming the calyx.
6. The Petals ; forming the corolla.
7. The Stamens ; bearing anthers which contain pollen-grains.
8. The Carpels. Ovary and ovules ; style ; stigma.
9. The ovary persists after the other parts of the flower fall away. The Fruit ; seed ; embryo and its parts.
10. A summary of the parts examined.

GATHER, first of all, a specimen of any common annual weed. It does not signify for our present purpose which you gather, provided it have a branching root, a distinct stem, and stalked leaves. Never mind the flower just at

present.   In gathering the specimen take it up carefully, so
that the root may be uninjured.

Proceed to examine your plant.

1. Observe the ROOT, noting in what respects it differs
from the parts which grow above ground.   It consists of
numerous fibres, or of a principal central root tapering
downwards and giving off irregularly many thread-like fibrils.
It is destitute of the green colouring of the stem and foliage,
being pale, or nearly white : it bears neither buds nor leaves,
and from its direction it appears to have avoided the light.

If you can find the tip of one of the root-fibres uninjured,

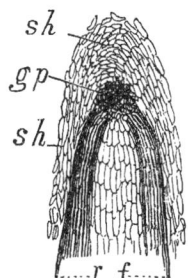

cut it off and examine it minutely with
your magnifying glass.   In case you
have not the means of examining it with
a high magnifier, you will find Fig. 1 a
sufficiently correct representation of it,
divided through the middle and magnified many times.

The point which I want you particularly to note is this :—The extremity of
the fibre is covered by a closely-fitting
sheath, protecting the actual growing
point, which is hidden immediately within the end of the sheath, to which it is
directly joined.   This protecting sheath
is being constantly renewed at its inner

FIG. 1. Longitudinal section through the ex-
tremity of a Root-fibre,
magnified. —*gp*, grow-
ing point ; *sh*, sheath ;
rf, root-fibre.

side by the "growing point ;" so that as the outer layers
become worn or withered by forcing a way through the
soil and pebbles, they are constantly replaced by inner
layers which take their turn, replace them, and then die, to
be in like manner replaced by fresh inner layers derived
from the "growing point" so long as the root continues to
live and to lengthen.

You will find then that the root avoids the light ; that it is pale or nearly white ; that its fibres give off irregularly numerous delicate thread-like branches (fibrils) ; that it is destitute of buds and leaves, and that the tips of the root are protected by cellular sheaths.

2. Now examine the STEM.

You observe, at once, that the stem rises above the ground, seeking rather than avoiding the light.   It may either be firm and erect, or weak and trailing.   Unlike the root, it is coloured more or less green, and not being usually woody, we may speak of it as *herbaceous.*   It bears several foliage-leaves arranged on different sides of the stem, the lower ones usually springing in a tuft from its base—at least while the plant is young.   The upper foliage-leaves are nearly always arranged singly or in pairs on the stem, although sometimes the only foliage-leaves of annual plants spring in a tuft apparently from the root.

If we examine the growing point of a young stem under a magnifying glass, carefully dissecting away the leaves which surround it, we shall find that to the very apex it continues to give off successively minute lateral prominences, which are the rudiments of leaves, either of foliage-leaves or of flower-leaves, for they both originate upon the stem in the same way, though they soon become different, both in their arrangement, form, texture, and colour.   In no case does the stem terminate in a cellular sheath like that which protects the tips of the root.

At the extremity of the principal stem of your plant, if fully grown, or upon certain of its branches, you find a tuft of coloured leaves forming a flower.

The branches spring from points where foliage-leaves are given off from the stem ; each branch occupying the angle (called the *axil* of the leaf) which the leaf makes with the stem. The branches may either resemble the primary stem from which they are given off, or they may differ in the kind of leaves which they bear, in the length of the intervals between their leaves, or otherwise.

The stem, we find, rises above ground; it is usually coloured green, and is herbaceous in texture; it bears foliage-leaves and flowers.

3. FOLIAGE-LEAVES. — I use the term foliage-leaves at present simply in order to avoid confusion with the leaves of which flowers are composed. It is not necessary you should always call them so, but it *is* necessary that you thoroughly understand that, speaking generally, whatever is borne by the stem and its branches is a *leaf* of *some kind*, whether it be green, as are foliage-leaves, or coloured, as are flower-leaves.

We have already remarked that the lowest leaves often *seem* to spring from the root. When this is the case these leaves may be called *radical* leaves. They really spring from a portion of the stem, which is more or less buried underground, giving off root-fibres below and radical-leaves from above. This portion of the stem in herbaceous plants which last more than a season or two is called the *stock*. The upper leaves, obviously springing from the stem, may be described as *cauline*.

The point on the stem from which any leaf is given off is called a *node;* the space between two nodes is called an *internode*. It is owing to the non-development of the internodes of the lower part of the stem or of the stock that the leaves which it bears appear to be given off in a tuft.

Each leaf consists of stalk and blade, or of blade only; the stalk supporting the blade is called a *petiole*. Leaves which consist of blade without any stalk are termed *sessile*. The same word is used of any part of a plant to denote the absence of a stalk, whether that stalk be a petiole (which is the stalk of a foliage-leaf only) or not.

The blade may be entire, or more or less divided into *segments*, or into separate pieces called *leaflets*. All the foliage-leaves have the blade spread out more or less horizontally, and they are all coloured green. They may be *hairy*, or *glabrous*, that is, destitute of hairs.

We find, then, the foliage-leaves to be borne by, and around, the stem ; they are thin, coloured green, and consist of petiole and blade, or of blade only : the blade being spread out horizontally. We now come to the examination of the—·

FIG. 2. Orange (*Citrus Aurantium*).

4. FLOWER ; and as the structure of the flower varies very much in different kinds of Indian annual weeds, it will be

needful for us to select some well-known plant which we can speak of by name, so that we may avoid misunderstanding.

Perhaps we cannot do better than take, first, the ORANGE-FLOWER, or the flower of one of its near allies, the Citron, Shaddock, or Lemon. The structure of the flowers of the latter is so similar to that of the flowers of the Orange, that any of them will do. As the leaves of which the flower consists are smaller than foliage-leaves, and very different from them in form, colour, and arrangement, it will be necessary that you be very careful in your observations, making sure that you thoroughly understand every stage of your progress.

The flower, observe, is borne upon a very short branch, which serves as a flower-stalk, and which is distinguished as the *peduncle* of the flower. Before proceeding to dissect (to separate carefully into its pieces) a flower, select one that has just opened, and which has lost none of the parts which it possessed while still a *bud;* that is, before it expanded.

Observe, first, that all the coloured leaves which form the flower are apparently arranged upon the very extremity of

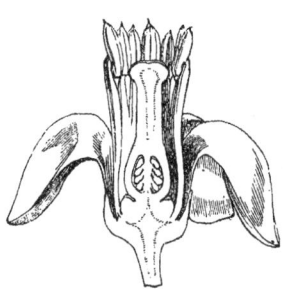

FIG. 3. Vertical section of the
Flower of the Orange.

the little branch which serves as a flower-stalk. The internodes which separate the upper foliage-leaves of the stem cease, or are *suppressed*, in the flower, so that all the parts are in close juxtaposition. This is characteristic of flowers. The top of the flower-bearing branch, from which the flower-leaves collectively spring, is called the *receptacle*, or *floral receptacle*.

5. Proceed next to note that the lowest and outermost part of the flower consists of a little cup with about five minute teeth upon its margin. These minute teeth indicate —as we shall become well assured as our experience widens —the number of leaves which are united to form the cup, each tooth answering to the tip of one of the little cohering leaves. These leaves are singly termed the Calyx-leaves, or *Sepals*, and together they form the *Calyx* of the flower, whether united, as in the Orange, or wholly separate, as in many common plants at hand in every garden. In the Orange the calyx remains after flowering, and may still be found even when the fruit is ripe. Such a calyx is said to be *persistent*.

We shall find it convenient to distinguish between a calyx which consists of separate calyx-leaves or sepals, and a calyx in which they are united more or less, terming the former *polysepalous*,[1] the latter *gamosepalous*. The calyx of the Orange is gamosepalous.

6. Immediately inside the calyx are normally five erect or slightly curved-back, separate, white, and wax-like leaves. These also are arranged in a whorl, and they are singly placed opposite to the intervals between the teeth of the calyx, not opposite to the teeth (sepals) themselves. Singly, they are the *Corolla-leaves* or *Petals;* together they form the *Corolla* of the flower.   The petals being free, the corolla is *polypetalous;* being equal in size and form, it is also *regular*.

Unlike the calyx, the corolla falls away early, and hence may be described as *deciduous*.

---

1 This prefix (πολύς, many) is used when applied to *sepals* and *petals*, to denote that the sepals or petals are *free* rather than that they are actually *many* in number. Usually there are not more than three, four, or five sepals or petals in a flower.

7. In the examination of the rest of the flower much
nicety is required. Having stripped off the calyx and
petals singly, and laid them aside, proceed to the third
series of flower-leaves. These are very different in form and
structure from both sepals and petals. They consist each
of a lower stalk-like portion, bearing an upper, somewhat
thickened, oblong, and grooved head. This stalk is termed

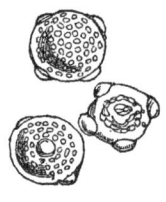

FIG. 4. Four Stamens of the Orange, three   FIG. 5. Three Pollen-grains of the
   of which are united by their filaments.   same.

the *filament*, the oblong head which it supports, the *anther;*
and the filament and anther together constitute a *Staminal
leaf* or *Stamen*. The stamens of the Orange are a little
shorter than the petals. As they are united to each other
by their filaments into sets of variable numbers, they are
said to be *polyadelphous*.

The anther we must examine more closely. We have
already observed that there is a groove up the *back* (outer
side), and another, less distinct, along the face (inner side).
These grooves divide the anther into two *lobes*, right and
left. If the anther be ripe, each of these lobes will split
open near the edge, allowing certain fine powdery granules
which it contains to be easily removed by insects or otherwise. These granules are essential to the flower as well as

to the stamen, so we must carefully examine them under a microscope. Fig. 5 shows them highly magnified. We find that they are distinct globular cells, with minutely granular contents. The fine powder is the *pollen*, and each of its globular cells is a *pollen-grain*.

Remove all the stamens, noting that they are, like the leaves of the corolla and calyx, inserted directly upon the floral receptacle, and immediately underneath the central organ of the flower. Such being the case, they are technically described as *hypogynous*.

8. You have now left in the centre of the flower, slightly raised upon a thickened cushion or ring, called the *disk*, the germ of the future fruit. If the lower part be cut across, you find it to be divided into a number of small cavities, radiating from the centre, each cavity containing the minute rudiments of future seeds. If you take the pains to count the number of cavities (you will probably find them about ten; they vary from seven to fifteen), you will ascertain the number of leaves which are united to form this central organ of the flower, which we shall speak of as the *pistil*. But how am I to convince you that the pistil is really composed of united leaves? It would not be easy to make this clear without reference to some other flower in which the pistil is of much simpler structure than in the orange-flower, so that I must ask you to lay it aside for a short interval while we examine the pistil of some pea-flower. Take the flower of the common Field or Garden-pea itself, or of a Scarlet-runner or Haricot, of Lablab, Chick-pea, Sunn-hemp, or Indigo. Any of them will serve, and one or another

FIG. 6. A longitudinal section of the Pistil of Orange, showing the interior of the ovary, with ovules, and style thickened upwards into the stigma.

cannot fail to be at hand.   Let us suppose that you have
gathered a flower of the Common Pea.   The other plants
with pea-flowers differ from it but slightly so far as the
pistil is concerned.

The calyx, you observe, is deeply five-toothed, indicating
that it is composed of five sepals, as in the orange.   As the
sepals are united below, the calyx is gamosepalous.   The
corolla consists of five petals, the uppermost of which is
much larger than the rest ; so that the corolla is said to be
irregular.   The same term is applied to any corolla or calyx
the parts of which are unequal in size or form.

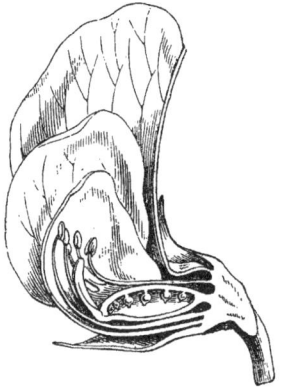

FIG. 7. Section of Flower of
Garden Pea.

FIG. 8. Diadelphous
Stamens of same.

Remove the petals carefully, especially the two lower
ones, which are united below by their edges into a boat-
shaped sheath called the *keel*.   The keel encloses the stamens, which in the Pea are, excepting one, united by their
filaments into a sheath, which is split open along the upper
side.   The single separate stamen is on the upper side of

the sheath next to the large petal.   Stamens united thus by
their filaments into two sets are said to be *diadelphous*.

Carefully cut away the stamens, and the central organ of
the flower, the pistil, remains, corresponding to the pistil of
the Orange.   Cut across with a sharp knife in its thickest
part, you find that it contains but a single cavity, and that
the minute rudiments of the future seeds are arranged in a
line upon the inner angle of the cavity.   These minute rudiments of the seeds are called, in their present stage, the
ovules, and the hollow leaf which contains and protects
them is called a carpellary leaf, or *carpel;* the lower,
hollow part, containing the ovules, being distinguished as
the *ovary*.

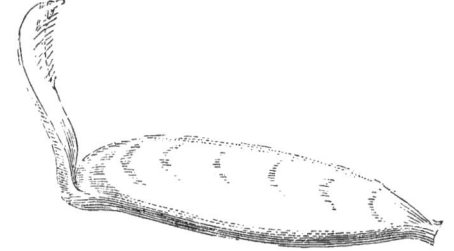

FIG. 9.   Pistil of Common Pea, enlarged.   The minute stigma occupies the
outer notch immediately within the apex of the bearded style.

The carpel is prolonged beyond the ovary into a long,
slightly flattened, up-curved tip, bearded with a row of short
hairs along its inner face.   At the oblique apex immediately
beyond the hairs is a minute naked projection, not discer-
nible without care, with a soft and cellular surface upon
one side.

This cellular surface is termed the *stigma*.   It is invariably present, and usually much more conspicuous than
it is in the Pea.   The stigma does not rest immediately

upon the top of the ovary in the pistil of the Pea, but it is
separated from it by a distinct stalk, abruptly narrowed
where it joins the ovary. This stalk supporting the stigma
is called the *style*. In many plants the stigma is without
a stalk, and rests upon the top of the ovary. When this
is the case, it is described as being *sessile*. In the Pea,
then, we have a pistil consisting of a single carpellary-leaf
answering to a foliage-leaf, a calyx-leaf,, a corolla-leaf or
a staminal-leaf, but differing from each of these in having
the margins curled in and united about half of its length,
so that it becomes hollow, and suited to protect the delicate
seed-buds (ovules) which are borne inside upon its united
margins.

Suppose, now, that instead of a single carpellary-leaf in
the flower, there were ten or fifteen of them arranged in a
close ring around the centre. The inner angle (that is, the
angle turned towards the centre of the flower) of each carpel
would correspond to the line of union of the edges of the
same carpel, and upon this inner angle the ovules would be
attached. If we suppose, further, all of these carpels to
cohere together into a single organ, we should have a pistil
similar to that of the orange-flower. The main difference
between the pistil of the Pea and that of the Orange is
simply this : that in the former the pistil consists of a single
carpel, in the latter of a number of cohering carpels. A
pistil consisting of a single carpel, or of two or more carpels
which do not cohere, is said to be *apocarpous*. A pistil consisting of two or more carpels cohering, though the extent
of their union be ever so slight, is said to be *syncarpous*.
The pistil of the Pea is apocarpous ; that of the Orange
is syncarpous. In each of them we have an ovary—in
the former one-celled, in the latter many-celled—style, and
stigma.

As the pistil in the Orange is wholly free from the calyx, it (or rather the ovary) is said to be *superior.*

9. Gather now another orange-flower; one in a more advanced state, with the petals and stamens all fallen away, and only the pistil remaining. The pistil here is passing into FRUIT. The style and stigma have withered up more or less, and probably fallen away altogether, leaving a scar on the top of the enlarged ovary, which has now become an " orange." In the ripe orange, as in the pistil at the time of flowering, the carpels continue to cohere, and their thin membranous sides divide the fruit into many *cells.* The principal change is due to the great increase in size of the ovary, to the formation of a quantity of juicy acid pulp in its cells, and to the development of the ovules into perfect *seeds,* one or more of which you may usually find in each cell of the fruit.

If we cut a ripe, plump seed right through lengthwise, we shall find that it consists of a tough, horny coat, the *testa,* enclosing one, or often two or more crumpled *embryos,* the germ of future orange-trees. It is so difficult to understand the structure of the embryo in the seed of the Orange, owing to the exceptional circumstance (which, however, is the rule in the Orange) of its containing more than one embryo, that we shall be obliged to go back to the Pea, or some other pea-flower, the seeds of which contain but a single embryo. Take a Pea, and, if hard, soften it by soaking it overnight, or by boiling it for a few seconds, so that you may strip off the testa enclosing the embryo. But before doing so, observe, first, the black spot on the side of the seed. This indicates the part by which the seed was attached to the fruit-carpel (*pericarp*) in which it was enclosed; it is the scar left on its separation from it. It is called the *hilum.* On careful examination you may observe at one end of the hilum a very minute aperture through the skin of the seed. You may find it by squeezing the soaked seed, when moisture issues from it. This is the *micropyle;* it answers to the micropyle of the ovule (see page 26). Generally, in ripe seeds, it is obliterated, or too minute to be observed.

As soon as the testa is removed, you find that its entire contents easily separate into halves, each half being plane on the inner face, and rounded on the outer. Observe, also, that these halves or lobes are hinged together at one side. Separate the lobes carefully, and you may observe upon the margin of the inner face of one of the lobes, close by the hinge, a rudimentary bud and root. You find, then, enclosed by the testa, (1) one pair of large seed-leaves ;

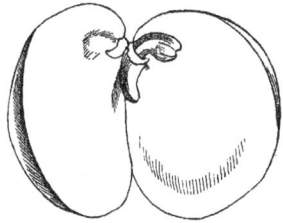
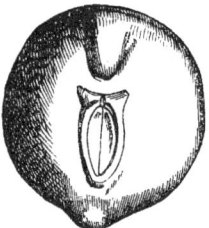

FIG. 10. Embryo of Pea with the cotyledons laid open, showing the curved plumule and radicle.

FIG. 11. Seed of same before removal of the testa, showing the hilum and ridge above indicating the position of the radicle inside.

(2) a bud with minute, rudimentary foliage-leaves ; and (3) the rudiment of a root. Nothing more. These parts are indicated in the cut, which shows the seed-leaves with their inner faces exposed. The seed-leaves are called *coty-ledons.* To the right is the bud of the stem, slightly curved inwards, called the *plumule;* and, pointing downwards, the rudiment of the root, called the *radicle:* the extremity of the

radicle invariably nestles immediately within the micropyle. The seed-leaves, or cotyledons, of the Pea are opposite; so we have an embryo with a *pair* of cotyledons, or a *Dicotyledonous* embryo.

Sow, if you please, a few peas, or a few orange-pips. Under favourable circumstances they will germinate, and grow up into plants similar to their parents.

10. Before we proceed to future chapters, in which we shall inquire into the relation of the various parts of the plant to each other, into the office or *function* which each is intended to perform, and the relation which the organs of other plants bear to the organs of the common species just examined, let us recapitulate the different parts which we have observed thus far. If there be any part which has not been clearly made out, make a point of understanding it before proceeding further.

We have, first, a ROOT, which descends into the soil, gives off fibrils irregularly, and is pale-coloured; the fibres have their extremities sheathed, and, as they do not give off the rudiments of leaves, the root is leafless. It is at first directly continuous with, and appears to pass into, the stem; but in plants which are fully grown, the original root is usually supplemented by other root-fibres which are given off from the bottom of the stem.

The STEM ascends, bears foliage-leaves, from the axils of some of which branches usually spring; it is coloured green more or less, and either itself terminates, or certain of its branches, in a tuft of coloured leaves forming the flower.

Root and stem, therefore, we find opposed to each other in the directions which they respectively take, as well as in several important points of structure. Together they may be regarded as constituting the *axis* of the plant; the root being the *descending*, the stem the *ascending* portion of the

axis.   Upon the ascending axis all the leaves, both foliage
and flower-leaves, are arranged.   The plant thus consists of
axis and *appendages.*

LEAVES.—These, we have found, are of five different
kinds.   First are the radical and cauline

(1) *Foliage-leaves,* called simply LEAVES.[1]

Then come the FLOWERS, terminating the stem or its
branches, consisting of—

(2) *Calyx-leaves,* called SEPALS.

(3) *Corolla-leaves,* called PETALS.

(4) *Staminal leaves,* called STAMENS.

(5) *Carpellary leaves,* called CARPELS.

---

[1] In many plants we may distinguish two modifications of the leaf
below the flower-leaves, besides the green foliage-leaves, but it is not
important to distinguish these at present.

CHAPTER II.

WHAT THE ROOT, STEM, AND LEAVES HAVE TO DO.

1. The plant fades. Why? Experiment shows that it is because water is withheld.
2. The root an absorbing organ.
3. Water is exhaled from the leaves. Transpiration. Absorption.
4. Other substances, besides water, are absorbed. The ash, and inorganic constituents of plants.
5. The organic compounds of carbon, oxygen, hydrogen, and nitrogen.
6. Ternary and quaternary compounds of these elements are in contrast.
7. Source of carbon in carbonic acid.
8. Liberation of oxygen by plants under the influence of sunlight. Assimilation.
9. Source of nitrogen.
10. Processes of absorption, transpiration, assimilation, and respiration, performed by "organs of nutrition," viz. the root, stem, and leaves.

1. By this time the weed which you gathered has probably faded; the leaves, now become soft and flaccid, are drooping, and the stem has lost much of its stiffness. How is this?

If the specimen be not quite withered, plant it again in the soil, and cover it with a flower-pot; or better and more easily done, put the root in water, and place the whole in a cool, shaded place for a few hours. We shall anticipate

C

matters and suppose that this has been already done, and
that you tried the experiment in this way upon three
distinct specimens. No. 1 you left lying upon the table.
No. 2 was placed with its root in water. No. 3 was hung
upside down, with a flower or leaf in water, the roots in
the air.

The general result of your experiment will be as follows :
—After the lapse of a few hours, No. 1 will be, as we have
already found, faded ; No. 2 will be nearly unaltered ; No. 3
will be partially faded, the parts out of water especially.
Hence we may gather that water supplied to the specimens
prevents them from fading, especially if it be supplied to
the root. On the other hand, if water be withheld, they
fade.

2. If we take now the faded specimen first described and
put its root in water, and leave it for a few hours in a cool,
shaded room, we shall probably find, unless it be irre-
trievably withered, that it freshens more or less ; the leaves
and stem become firmer and more nearly like their
original state.

This experiment shows us, further, that water supplied to
a fading plant enables it to recover.

Reflecting upon these experiments, we shall be led to
the following conclusions :—

    i. That water evaporates from the exposed surface of
        plants.

    ii. That fresh supplies are taken into the plant by the
        root.

    iii. That the stem serves to convey this water-supply from
        the root to the leaves.

3. We may now try another simple experiment, devised
by Professor Henslow, which shows that exposure to direct
sunlight, as well as dryness of the air, has to do with this

evaporation of water from the leaves.   Take six or eight
tolerably large, healthy leaves with petioles an inch or two
in length ; two tumblers filled to within an inch of the top
with water, two empty dry tumblers, and two pieces of
card, each large enough to cover the mouth of a tumbler.
In the middle of each card bore three or four small holes
just wide enough to allow the petiole of a leaf to pass
through.   Let the petioles hang sufficiently deep to dip into
the water when the cards are put upon the tumblers containing it.   Having arranged matters thus, turn the empty
tumblers upside down, over each card, so as to cover the
blades of the leaves.   Place one pair of tumblers in the
sunshine, the other in a shady place.   In five or ten minutes
examine the inverted tumblers.   That exposed to the sun
you will find already lined with dew on its cool side, while
that kept out of the sun is still nearly or quite clear.   It is
manifest, therefore, that evaporation from the leaves must
be not only rapid, but considerable in amount when plants
are exposed to the sun or a dry atmosphere.

This exhalation of vapour from the surface of plants is
termed *transpiration*.   A correct understanding of the process explains how it is that plants growing in parlours are
apt to become faded even when watered, because the taking
up of water (termed *absorption*) by the roots cannot keep
pace with the transpiration from the leaves, owing to the
rapid evaporation excited by the dryness of the air.   Since
the specimen No. 3, experimented upon at the beginning of
the present lesson, faded, notwithstanding the immersion of
some of its leaves, it is clear that the root is the part which
performs the office of absorption principally.

Every part of a plant or animal appropriated to a distinct
purpose or function is termed an *organ*.   Hence the root
may be called the *organ of absorption* of the plant.

4. Now, not only is water absorbed by the root, but also various substances which are dissolved in the water. Hence we find, if we burn a plant carefully, that an *ash* remains, consisting of such of these substances as are not dissipated by heat, which were absorbed in this way, and which had been made use of by the plant, or stored away in its tissues. Of the simple elements known to chemists, about twenty occur in the ash of the plants ; many of these, however, in very minute quantities, and never all in the same plant. Sulphur, phosphorus, potash or soda, lime and silex, are those most generally found.

5. But if we analyse an entire plant, and not the ash only, we shall find constantly present, besides the above, the elements carbon, oxygen, hydrogen, and nitrogen. And these elements are present, there is reason to believe, in every organized being, whether plant or animal, in combinations peculiar to organized beings. Hence they may be called the *organic elements,* in contradistinction to the (mineral) elements found in combinations which are not peculiar to organized beings, and several of which remain in the ash of plants when burnt. The latter may be called the *inorganic elements.*

6. These four organic elements do not exist separately in the plant, but, as we have said, in combination. Thus the carbon is united with oxygen and hydrogen (the two last the elements of water), forming the basis of a series of compounds, called *ternary compounds* because they are composed of three elements. The nitrogen occurs combined with the same three elements, forming a *quaternary compound,* or compound of four elements. And these two series of organic compounds stand in remarkable contrast to each other in the plant, both in respect of the structures in which they respectively take part, and of function, as we

shall point out when we come to speak of the minute structure of plants.

7. We have already explained how water (oxygen and hydrogen) finds access to the plant, as well as the mineral substances which may be held in solution by the water. With regard to the important element carbon, experiments clearly show that it is absorbed in combination with oxygen, as carbonic acid gas, which is soluble in water, and may thus be taken up by the root. Carbonic acid gas also occurs in the atmosphere, and green leaves, under the influence of direct sunlight, possess the power of absorbing it directly from the air.

8. But the most remarkable circumstance attending this absorption of carbonic acid is the liberation of oxygen gas by the leaves, very nearly to the amount absorbed in combination with the carbon of the carbonic acid gas. This liberation of oxygen is most easily shown by taking a few leaves which have been first soaked a day or two in water so as to become saturated, and exposing them, plunged in water containing carbonic acid (as ordinary spring or pump water, in which it is always present), to direct sunlight. Minute bubbles will be given off, under favourable circumstances, in a rapid and continuous stream. These bubbles consist of almost pure oxygen.

This fixation of the carbon and liberation of the oxygen of carbonic acid has been termed vegetable *respiration*, but as the conditions which obtain are the reverse of those characteristic of animal respiration, it may be more correctly spoken of as a part of the process of vegetable *assimilation*.

Repeated experiments have shown that some of the parts of the flower, seeds when germinating, and also plants which are not coloured green, absorb oxygen from the air, and give off carbonic acid gas. This may be regarded as a respiratory process. It is not improbable that the green parts of plants also may at all times, but especially in the dark, absorb oxygen and give off carbonic acid in the same way, but in these parts the amount of carbon fixed greatly predominates over that which is liberated. The mutual relations, however, of these processes are as yet very imperfectly understood.

9. It is not yet perfectly clear from what immediate source the plant obtains its nitrogen; not that the element is scarce, since it forms four-fifths of the atmosphere, but the precise state in which it enters the plant, whether uncombined (which is not probable) or in combination, as in ammonia and nitrates, is still a matter of inquiry and discussion among scientific men.

10. The processes which we have briefly described of *absorption, transpiration, assimilation,* and *respiration,* we find mainly concern the root, the stem, and the leaves. These are the food-providers and preparers of the plant. Hence we call them collectively the *nutritive organs;* the root being, as we have shown, the organ of absorption, the foliage-leaves specially of transpiration, assimilation, and respiration. The stem, when green, assists the leaves in their work; but, speaking generally, it may be regarded merely as the support of the leaves, maintaining their connexion with the root.

In our next chapter we shall inquire into the mutual relations and functions of the leaves which compose the flower, deferring further reference to the chemistry of the organs until we speak of their minute structure.

CHAPTER III.

1. WE have already seen in the case of our Orange, that the flower results in a fruit, each division of which, answering to a carpel, usually contains one or more seeds. The seed we found to contain the minute germ of a future Orange plant, which we called the embryo. As it is the special function of all the leaves which compose the flower to contribute to this formation of embryo-containing seeds, by means of which the Orange is enabled to reproduce and multiply its kind, we may term all the parts of the flower *Organs of Reproduction*, in contradistinction to the organs considered in our last chapter, which contribute primarily to the conservation of the individual Orange-tree, and which,

from their functions, are styled, collectively, *Organs of Nutrition.*

2. The four series of leaves of which the flower is composed do not each fulfil the same part in the production of the seed. The corolla and the stamens are deciduous. They fall away, leaving the pistil to mature into fruit. But they do not generally fall until after an important function has been accomplished by the stamens, either of the same flower, or of another Orange-flower.

The two outer series of flower-leaves, the sepals and petals, may be regarded simply as organs designed to protect the smaller and delicate parts which they enclose during their early development ; and perhaps, also, the colour and greater size of the petals may serve to attract insects which, we shall find, have an important work to perform, as aids in securing the formation of good seed. Hence the calyx and corolla are termed the *envelopes* of the flower. As both calyx and corolla are present in the Orange-flower, the envelope of the flower is *double,* or in two series. Hence the flower is termed *dichlamydeous.*

3. The anthers we have observed are divided lengthwise into two lobes, which lobes, after the expansion of the flower, become fissured near their margins, so as to liberate the grains of pollen which they contain. About the time that the anthers open to discharge their pollen, we may observe that the stigma is rough, with microscopic projecting cells which, upon minute examination, we shall find to be slightly moistened. Upon the stigma, after the flower has been open for a few hours in fine weather, there may usually be found a few grains of pollen, which have either reached the stigma by direct contact of the anthers, or by means of some insect visiting the flower in search of honey, and which, unwittingly, conveyed some of the pollen, acciden-

tally adhering to its hairy limbs and body, to the stigma. This transfer of the pollen from the anther to the stigma is highly important. If we separate a few stamens, with their anthers and pollen, and keep them apart from the rest of the flower, or if we remove the pistil in the bud so that stamens only remain within the envelopes, we shall find that they ultimately shrivel and wither up, pollen-grains and all, without undergoing further change. But the case is different with the grains of pollen which reach the stigma. After an interval, varying in different plants from a few hours to several months, we find the pollen-grains begin to grow, and their growth takes place in this way :—

Each grain of pollen, as we have already learnt, is a single cell. These cells almost invariably have a *double* coat, an outer and an inner; and in the outer coat there are frequently thin places, or actual openings here and there,

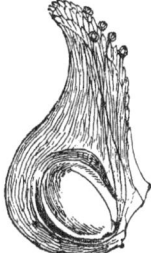 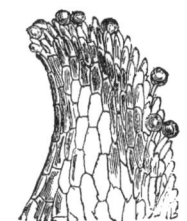

FIG. 12 *a*. Diagram representing Pollen-grains upon the stigma of a carpel of Ranunculus, which have developed their tubes, reaching to the micropyle of the ovary. The tubes are so delicate that it is impossible to trace them the whole way.

FIG. 12 *b*. The Stigma magnified, with grains of pollen upon it.

which permit the inner coat to grow through them at one or more points. This growth of the inner coat of the pollen-grain is encouraged by the moisture which bathes the stigma,

so that at length it protrudes, and, like an excessively minute
root-fibre, penetrates the substance of the stigma, and passes
down through the short style until it reaches the cavity
of the ovary.   As the changes of which we speak can only
be observed under a considerable magnifying power, we
shall explain them more clearly by reference to the cut, Fig.
12*a*, which represents some grains of pollen which have
developed tubes reaching into the ovary.

4. The ovary contains several minute seed-buds, the
ovules; which ovules in the Orange are inverted (*anatropous*).   Each ovule consists of a central cone, called
the *nucleus* of the ovule, around which central cone is a
layer of cells forming the *coat* of the ovule.   This cellular
coat grows up around the nucleus, and closes over it
excepting at the top, where a very minute aperture through
the coat is always left.   This aperture is called the *micropyle.*   Owing to the ovules of the Orange being anatropous,
the micropyle is brought close to their point of attachment
(*hilum*); and as they are pendulous more or less, it is
directed upwards.

By the time that the pollen-tube has reached the cavity
of the ovary, certain important changes have taken place
in the cells which form the nucleus of each ovule.   One
cell has enlarged greatly at the expense of its neighbours, so
as to occupy a considerable part of the nucleus.   This
enlarged cell is called the *embryo-sac*, because within it we
find the embryo to originate.   But this is not until after
the pollen-tube has reached the micropyle of the ovule and
actually penetrated to the upper end of the embryo-sac,
against which it becomes closely applied.   Presently, after
this contact of pollen-tube and embryo-sac, a cell forms
within the latter, which ultimately developes into the
embryo, and the ovule then becomes the young seed.

5. We can now understand why the petals and stamens are deciduous. Their function is soon accomplished, and their texture is too delicate to allow them to persist, although the calyx of the Orange, from its firmer texture, remains until the fruit becomes ripe. A calyx or corolla remaining thus attached after the time of flowering is said to be *persistent*.

6. From what we have here described of the functions of the different organs of the flower, the high importance to the plant of their proper performance must be plain. And from the general constancy which the parts of flowers present in their principal features, both in the structure of the several parts and in their relations to each other, in groups of plants which from numerous general resemblances we may reasonably imagine to be *related by descent* (that is, related to each other in the same way that Europeans are more nearly related to each other than they are to the Negro or Indian races, or as fish are more nearly related to each other than they are to birds or reptiles), botanists make use of characters afforded by the organs of the flower and fruit to mark, in words, the principal divisions of the vegetable kingdom. Hence it is desirable, before we proceed to consider the organs which are more subject to variation, that we should acquire a correct notion of the nature of the principal modifications to which the parts of the flower are liable in different plants.

With a view to this, and that you may be enabled at once to commence the examination and description of flowers, we shall proceed in our next to compare, with those of the Orange and Pea, the flowers of a few common plants representing the most important types or kinds of modification of floral structure.

CHAPTER IV.

1. GATHER flowers of as many of the following common plants as you are able. The accompanying woodcuts must do duty for those which are not in flower, or which happen to be out of reach :—

Opium Poppy; Indian Mustard or Rape ; any Rose with single flowers ; Melastoma Malabathricum ; Garden Zinnia ; Rose Periwinkle (*Vinca rosea*) ; Basil (*Ocymum Basilicum* or *O. Sanctum,* known as Tulsee) ; Grass-cloth Nettle (*Bœhmeria nivea*) ; Willow ; Colocasia Antiquorum, known as Kuchoo, Kachalu, Ghwian, or Kandalla ; Garden Dendrobe (*Dendrobium nobile*); any Dracæna; Crinum asiaticum, known as Buro-kanoor, Sukh-dursan, or Tolabo ; Wheat.

There are three conditions which play a most important part in modifying the structure of flowers, to which we must direct attention before proceeding.   These are *cohesion, adhesion,* and *suppression.*   The first two terms are used by botanists to denote the union of like (cohesion) or of unlike (adhesion) parts of the flower.   Thus union of sepals with sepals, of petals with petals, of stamen with stamen, of carpel with carpel, is said to be due to *cohesion,* parts of the same whorl or series being concerned.   Union of corolla to stamens, or of ovary to calyx, or of stamens to calyx or corolla or to the pistil, is due to *adhesion,* parts of different whorls or series being concerned.

The term *suppression* is used to denote the absence of parts in a flower which, from analogy, we might expect to find.   Extended observation shows that the number of sepals, of petals, and of stamens, is, in a large proportion of flowers, the same ; or the stamens may be a multiple— twice or three times as numerous, for example, as the petals or sepals.   Thus, when we find that in some flowers the corolla is absent, in others the corolla and stamens, or the corolla and pistil, we speak of such parts of the flower as being suppressed.   That this is generally a correct view to take with respect to the absence of organs we find confirmatory evidence in the frequent imperfect or partial development of such organs in plants allied in other points of structure to those in which they are wholly absent. Sometimes but a single series of organs, either stamens or pistil, constitutes the flower, the three other series being suppressed.   Single parts, also, of a series, as a sepal, a petal, &c., when absent, are said to be suppressed.

In the Orange-flower we have neither adhesion nor suppression of parts, but we have cohesion of sepals, of stamens, and of carpels.   So that it may be described as with—

> Calyx *inferior, gamosepalous (persistent).*
> Corolla *hypogynous, polypetalous.*
> Stamens *hypogynous, polyadelphous.*
> Pistil *syncarpous,* ovary *superior.*

In the examination of the flowers just enumerated we shall find manifold variety in respect to these conditions of cohesion, adhesion, and suppression.

2. POPPY (*Papaver*).—Either the white-flowered Opium Poppy, or any single scarlet Poppy, or the common introduced weed, the yellow-flowered Mexican Horned-poppy, will serve.   It will be needful to gather a bud as well as an expanded flower, whichever may be chosen, because the calyx is thrown off very early by the expansion of the large crumpled petals.   Organs thrown off

FIG. 13. Opium Poppy (*Papaver somniferum*).

thus at or before the time of expansion of the flower are

said to be *caducous*. You find the calyx consists of two,
or sometimes of three, distinct sepals, inserted outside of
and below the other organs of the flower. The corolla
consists of four or six free petals. The stamens are very
numerous and free. The pistil, like that of the Orange,
consists of numerous cohering carpels, as indicated by the
lobes or rays of the stigma. In the Horned-poppy the
coherent carpels are four to six in number, while in the
true Poppies they are indefinite. If the ovary be cut across,
there will be found as many projecting plates or lines bearing ovules as there are stigmas, so that the syncarpous
character of the pistil cannot be doubted. The Poppy may
be described as with—

> Calyx *inferior*, and *polysepalous (caducous)*.
> Corolla *hypogynous* and *polypetalous*.
> Stamens *hypogynous* and *polyandrous*.
> Pistil *syncarpous*, ovary *superior*.

3. INDIAN MUSTARD, or RAPE (*Brassica*).—Either will
do. There are four free sepals, four free equal petals and
six free stamens, of which four are long and two short
(hence called *tetradynamous*). The slightly two-lobed stigma
indicates the syncarpous condition of the pistil, which is
regarded as consisting of, at least, two coherent carpels.
The flower of Mustard or Rape may be described—

> Calyx *inferior*, *polysepalous*.
> Corolla *hypogynous*, *polypetalous*, *regular*.
> Stamens *hypogynous*, *tetradynamous*.
> Pistil *syncarpous*, ovary *superior*.

4. ROSE (*Rosa*).—Any single-flowered rose will serve.
The calyx at first sight appears to consist of five distinct

FIG. 14. Indian Mustard (*Brassica juncea*).

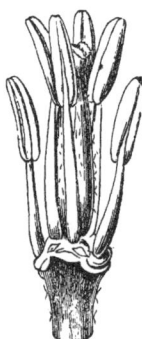

FIG. 15. Tetradynamous Stamens of cruciferous flower.

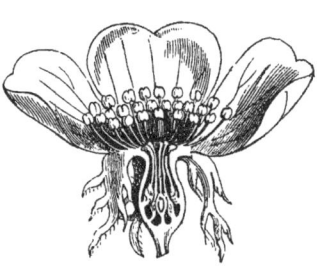

FIG. 16. Rose (*Rosa*). Vertical section of a flower.

sepals, but if the flower be cut vertically from below upwards through the middle, you will observe that the five apparent sepals are inserted upon the margin of a flask-shaped cavity, upon the inner surface of which several carpels are borne. This cavity is usually regarded as the *tube* of the calyx, of which the five lobes surrounding it constitute the *limb*, and indicate, at the same time, the number of sepals composing the calyx.   Five equal petals and numerous stamens are inserted in the mouth of the tube, and lower down in the tube are the numerous distinct carpels composing the pistil of the flower, with their styles closely crowded together in the narrow throat of the calyx-tube.   The corolla and stamens being inserted upon the calyx and not upon the floral receptacle immediately below the pistil, as in the flowers just examined, they are said to be *perigynous*.   Carefully note that the carpels are wholly free from the tube of the calyx, although the latter rises nearly, or quite, to the level of the stigmas.   Describe the Rose as with—

> Calyx *inferior*, *gamosepalous*.
> Corolla *perigynous*, *polypetalous*, and *regular*.
> Stamens *perigynous*, *indefinite*.
> Pistil *apocarpous*, ovaries *superior*.

5. MELASTOMA MALABATHRICUM.—The calyx is composed of five cohering sepals, as indicated by the five lobes of its limb.   Five equal and distinct petals and ten curious stamens are inserted upon the calyx.   Five of the stamens are alternate with the petals, and five are opposite to them. A transverse section of an ovary will show that the pistil is syncarpous, consisting of five carpels, as indicated by the five cells of the ovary.   The style is undivided.   If a flower be cut through vertically, you will observe that the ovary adheres at intervals to the calyx-tube, leaving pocket-like

interstices into which the anthers are packed before the flower expands.   As this adhesion to the calyx does not

Fig. 17. Melastoma Malabathricum.   Fig. 18. Flower of Melastoma.

extend to the top of the ovary, the latter may be described as half-superior.

> Calyx *half-inferior, gamosepalous.*
> Corolla *perigynous, polypetalous.*
> Stamens *perigynous, decandrous.*
> Pistil *syncarpous*, ovary *half-superior.*

6. GARDEN ZINNIA (*Zinnia elegans*).—The structure of its flower-heads is very puzzling to beginners.   The best way to understand it is to make a vertical section right through the middle of one of the heads with a sharp knife, cutting from below upwards.   You will then find that what appeared at first to be a single flower with spreading richly-

coloured petals, is in reality a head of numerous flowers (*flower-head*). The flower-head is surrounded by a number of scaly leaves, forming an *involucre*, called by the older botanists a " common calyx." Within the involucre the summit of the stem rises in a conical form, bearing, closely packed upon its surface, the little flowers, called *florets*, of

FIG. 19. Garden Zinnia (*Zinnia elegans*).

the flower-head. There is a marked difference in form between the outer and inner florets of the flower-head, due to the one-sided enlargement of the corolla in the former.[1] The outer florets with one-sided corollas, taken together, form the *ray* of the flower-head ; the smaller florets, with

[1] In a recent variety of the *Zinnia* the flower-heads show a tendency to become " double," by all the florets acquiring the form of corolla, which in the wild state and common varieties is characteristic of the ray-florets only.

regular corollas, occupying the centre of the head, form the *disk.*

In describing the structure of flower-heads (*capitula*), it is well to examine the ray and disk florets separately. Neither of these appear, at first sight, to have a calyx. Analogy,

FIG. 20. Vertical section of a capitulum of Garden Zinnia, showing two ray-florets and one disk-floret.

however, affords sufficient reason to conclude that each floret has its own calyx, but it is wholly adherent to the ovary. It is superior and gamosepalous. In many plants related to the Zinnia—the Thistle, for example —the upper free portion (*limb*) of the calyx exists as a crown of fine bristles surrounding the top of the ovary. The corolla of the ray-florets is gamopetalous and irregular; of the disk-florets, gamopetalous and regular. In both it is inserted apparently upon the top of the ovary, and is consequently termed *epigynous.* The stamens in the ray-florets are absent or imperfect; in the disk-florets they are five in number (pentandrous), and inserted upon the corolla. In consequence of this adhesion to the corolla they are termed *epipetalous.*

FIG. 21. Fruit of Thistle, with sessile, plumose pappus.

An important character which the stamens present is yet to be noted.   If the tube of the corolla of one of the disk-florets be very carefully laid open with the point of a pen-knife, it will be found that the five stamens cohere by their anthers.   On this account they are termed *syngenesious.*

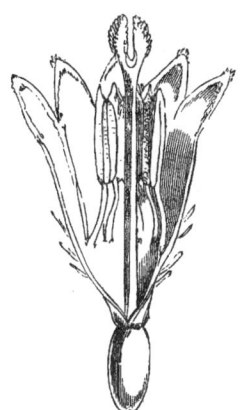

FIG. 22.   Tubular floret of Composite laid open, showing the syngenesious anthers.

The coherent anthers form a tubular sheath closely surrounding the style.   The pistil we may infer to be syncarpous, from its two-lobed stigma, notwithstanding that the inferior ovary is one-celled.

In Zinnia you find the—

> Calyx *superior, gamosepalous.*
> Corolla *epigynous, gamopetalous.*
> Stamens *epipetalous, pentandrous, syngenesious.*
> Pistil *syncarpous,* ovary *inferior.*

7. ROSE PERIWINKLE (*Vinca rosea*).—The calyx is free, but the five sepals are coherent below.   The corolla is

hypogynous and regular, consisting of five cohering petals, as indicated by the five spreading divisions of the limb.   The five stamens are inserted upon the tube of the corolla, and

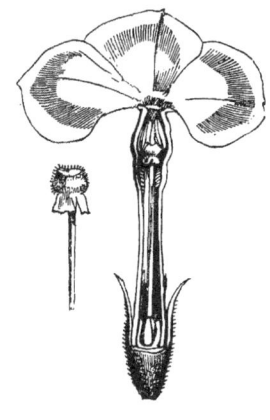

FIG. 23. Rose Periwinkle (*Vinca rosea*).

FIG. 24. Flower of Rose Periwinkle laid open.   To the left the stigma and portion of style.

are consequently epipetalous.   Stripping away the calyx and corolla, the pistil will be found consisting of two carpels, which are separate below, but coherent above, so that there are two ovaries and but a single style and stigma.   Rose Periwinkle may be described—

> Calyx *inferior, gamosepalous.*
> Corolla *hypogynous, gamopetalous.*
> Stamens *epipetalous, pentandrous.*
> Pistil *syncarpous,* ovary *superior.*

8. BASIL (*Ocymum Basilicum* or *O. Sanctum,* known as Tulsee).—Five coherent sepals form a gamosepalous calyx,

which is free and inferior ; five coherent petals form a gamopetalous corolla, which is irregular and two-lipped (*bilabiate*). The number of stamens falls short of the petals, there
being but four ; so that we may conclude one stamen is suppressed, since in Rose Periwinkle, and in most plants with
regular gamopetalous corollas, the number of stamens equals
the number of petals cohering to form the corolla. It is
worth noting that a tendency to irregularity in gamopetalous
corollas is often accompanied by a tendency to suppression
of one or more stamens. In some species we find the

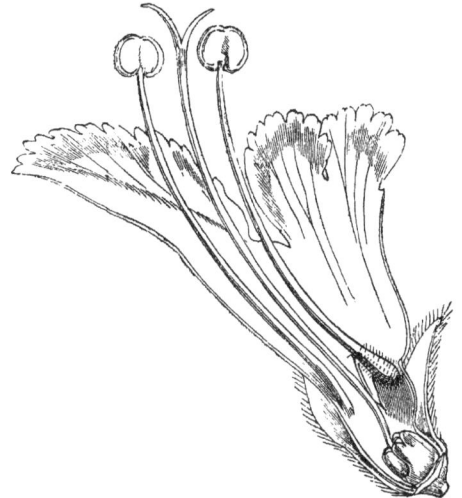

Fig. 25 Vertical section of Flower of Basil.

missing stamen imperfectly developed, confirming the cor-
rectness of this view. The four stamens are not equal in
length, one pair being rather longer than the other. On
this account they are said to be *didynamous.* Observe,

also, that the stamens are adherent to the lower part of the corolla. They are consequently *epipetalous.* The pistil is syncarpous, consisting of two carpels, as indicated by the bifid stigma, and the ovary is superior and deeply four-lobed, so that the style springs from the centre and base of the lobes of the ovary. Basil has—

> Calyx *inferior, gamosepalous.*
> Corolla *hypogynous, gamopetalous, irregular.*
> Stamens *epipetalous, didynamous.*
> Pistil *syncarpous,* ovary *superior.*

9. GRASS-CLOTH NETTLE (*Bœhmeria nivea*).—It will be needful to gather two specimens, carefully examining the minute flowers, in order to ascertain that in one specimen they enclose stamens, in the other a pistil, because these organs occur not only in separate flowers, but, in this plant, often upon distinct individuals. Flowers such as these, con-

FIG. 26. Staminate Flower of Grass-cloth Nettle (*Bœhmeria nivea*). 　　　FIG. 27. Pistillate Flower of same.

taining stamens only, or pistil only, are said to be *imperfect, unisexual,* or *diclinous.* When both staminate and pistillate (unisexual) flowers occur upon one and the same plant, they are said to be *monœcious ;* when, as in Grass-cloth Nettle, upon distinct individuals, they are said to be *diœcious.*

Examine the staminate and pistillate flowers separately. In the staminate flower you observe the calyx to consist of four sepals, which are slightly coherent at the base. As the corolla is suppressed, the envelope of the flower is single (*monochlamgdeous*), not double (*dichlamydeous*), as in the species previously examined. Opposite to the sepals are the four stamens inserted upon the receptacle. The pistil is represented by a minute central rudiment.

In the pistillate flower the calyx consists, also, of four leaves; but here they are coherent nearly to their tips, forming a flask-like calyx-tube, which closely invests the ovary, without, however, adhering to it. Four minute, unequal teeth, indicate the number of sepals cohering to form the calyx. There are no stamens, and the pistil consists of a single carpel, with a superior ovary.

In Grass-cloth Nettle we have the flowers :—

Calyx *gamosepalous, inferior.*
Corolla o.
In the male flower (♂), Stamens *hypogynous, tetrandrous.*
Pistil o.
In the female flower (♀), Stamens o.
Pistil *superior, apocarpous.*

10. WILLOW (*Salix*).—It does not matter as to the species, but, as in the case of the Grass-cloth Nettle, two specimens, from different trees, will be required, as the flowers are diclinous and dioecious. You find the flowers arranged in spikes, which, being deciduous and bearing imperfect flowers, are specially distinguished as *catkins*. Both the staminate and pistillate flowers are destitute of calyx and corolla. Having therefore no envelope to the essential organs, they are called *achlamydeous*. The stamens, two, three, five, or rarely. more, spring from the axil of a minute

scale-like leaf (bract), and constitute the male flower, of
which a number are crowded together upon the same catkin.
The pistil of the female flowers also springs from the axil of
a similar bract; it is syncarpous, consisting of two carpels,

FIG. 28. Staminate Flower of Willow.   FIG. 29. Pistillate Flower of same.

as indicated by the bifid stigma and two short lines of
ovules in the single cavity of the ovary.   The flowers of the
Willow figured above may be described thus :—

Calyx o.

Corolla o.

Male flower—Stamens 2 (*diandrous*).

Pistil o.

Female flower—Stamens o.

Pistil *syncarpous*.

In the Weeping Willow (*Salix babylonica*), much planted
in India, the staminate flowers are usually diandrous; in the
common indigenous Willow (*S. tetrasperma*) they vary six
to eight.

11. As the plants which we have hitherto examined differ
in many important particulars from those which yet remain
of the fourteen enumerated at the beginning of the chapter, it

may be well to review here some general characters afforded both by the reproductive and nutritive organs, which are common to all those we have already done, and which are more or less markedly in contrast with the characters presented by the corresponding organs of the plants yet to be examined.

In nearly all the plants examined you find the leaves with a distinct blade and petiole ; and, if you hold the blade of any of them up to the light, you may notice that the small *veins* which ramify through it are netted *irregularly*. In the flowers, you have observed that the parts of the calyx (sepals) and of the corolla (petals), whether free, coherent, or adherent, are either in *fours* or *fives ;* that is, four or five to a whorl.

12. Now the characters of (1) leaves more or less distinctly narrowed at the base into a petiole ; of (2) irregularly net-veined leaves ; and (3) the arrangement of the parts of the flower in fours or fives (which three characters we have found to apply more or less to all the specimens which we have examined hitherto), are supported by other characters afforded by the seeds and mode of growth of the wood, which it is important you should correctly understand.

13. If we put a few peas upon moist earth in a flower-pot and cover them with a bell-glass, the first stage of growth, termed *germination,* of the young pea-plant may be conveniently observed. The essentials to germination are found by experience to be a certain amount of moisture, warmth, and air. If sufficiently warm (and the amount of warmth required to commence with varies in the seeds of different plants), moisture is absorbed by the seed, which causes it to swell up so as to burst the seed-skin. Oxygen also is absorbed from the air, and certain chemical changes, accompanied with the liberation of some carbonic acid, take place in the cells of the embryo, resulting in the solid

substances which they contain being made available for the use of the growing plant. The radicle is always the first to break out, curving down towards the earth, whatever may be its position. The radicle, by its direct prolongation, forms the primary root of the plant. The plumule shortly after disengages itself, ascends and developes into the stem of the Pea, bearing foliage and flower-leaves.

14. We have already briefly described the structure of the seed of the Pea, which we found to consist of an embryo enclosed by the testa, or seed-coat. Take now, for comparison with the seeds of the Pea, a few fresh seeds of the Castor-oil plant. The hilum or scar indicating the part at

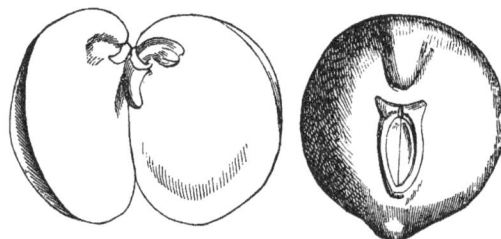

Fig. 30. Seed of Pea ; to the left the embryo, with the cotyledons laid open.

which the seed was attached in the cavity of the fruit is at the top of the seed, and the micropyle, which it is difficult to discover in the ripe seed, is close to it. Remove the thin shell-like and prettily mottled testa, and also the thin membranous inner covering of the seed, and at the top, immediately under the micropyle, you may find the point of the radicle of the embryo directed upwards. Cut the seed through lengthwise and at right angles to its greatest breadth, beginning at the radicle, and you will find its parts as represented in the accompanying cut (Fig. 31). An embryo, con-

sisting of two large thin cotyledons, and a radicle, enclosed in
a quantity of a uniform, white, firm substance, resembling the
cotyledons of the Pea in texture.   This substance, in which
the embryo is embedded, is called the *albumen* of the seed.
It is at the expense of this albumen that the embryo is
enabled, during its germination, to develope a root and stem.
The albumen in the seed of Castor-oil substitutes the store
of nutrient matter contained in the thick cotyledons of Pea,
the embryo alone of which, with the testa, constitutes the
entire seed.   Seeds containing, besides an embryo, a deposit
of albumen, whether large or small, are said to be *albuminous*.   Seeds like those of the Pea and Orange, which
contain an embryo only, are *exalbuminous*.   Between the

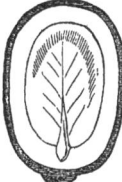 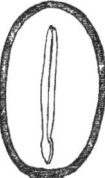

Fig. 31. Vertical sections ot seed of Castor-oil plant (*Ricinus communis*) ; to the
left cut in the plane of the cotyledons, to the right at right angles to the
cotyledons, showing the copious albumen in which the embryo is embedded.

two extremes of abundant albumen (Castor-oil plant) and no
albumen at all (Pea, Orange) we have every degree.

15. Like the Pea, therefore, the Castor-oil plant is dicotyledonous ; and as the character expressed by this term (the
possession of a pair of cotyledons) is common to plants with
irregularly net-veined leaves, and with the parts of their
flowers in fours or fives (with but a comparatively small
number of exceptions), botanists employ the term DICOTY-
LEDONS as the name of a great *Class* of flowering plants,
including all those which present the above characters.

16. It must always be borne in mind, however, that none of these characters are absolute. They are always subject to exception. So that plants which exhibit a departure from the prevalent type of Dicotyledons in any single character only are still referred to the same Class. Thus we have a few Dicotyledons which are actually destitute of cotyledons, or which have but one, or more than two ; we have some with parallel-veined leaves, and others with the parts of the flower in threes. But in all these cases the question as to which Class the plant shall be referred, is decided, not by any solitary character, but by the sum or preponderance of characters which it presents.

The structure of the stem and mode of growth of the wood we shall speak of in a later chapter.

CHAPTER V.

1. Colocasia antiquorum (known as Kuchoo, Kachalu, Ghwian, or Kandalla).—Without much care you will be liable to misunderstand the structure of this plant, as did Linnæus himself that of a near European ally. The flowers are closely packed in rings upon the lower part of the fleshy spike, which you find enclosed in a large sheathing

bract-leaf, called a *spathe*.   A flower-spike of this kind, enclosed in a spathe, is distinguished as a *spadix*.   With a magnifying-glass compare the structure of the minute flowers

FIG. 32. *Colocasia antiquorum.*

occupying the lower part of the spike, which are similar to the single detached flower to the left in the cut (Fig. 34), with those a little higher up, which are similar to the de- tached flower to the right in the cut.   Be careful to note, however, that between the two belts of different flowers there occurs a short, often slightly narrower, intermediate portion of the spadix occupied by a number of rudimentary organs which must not be mistaken for efficient flowers.

The flowers of the lower belt consist each of a pistil only, each pistil composed of two or three united carpels, as indicated by the slight lobing of the cellular surface of the

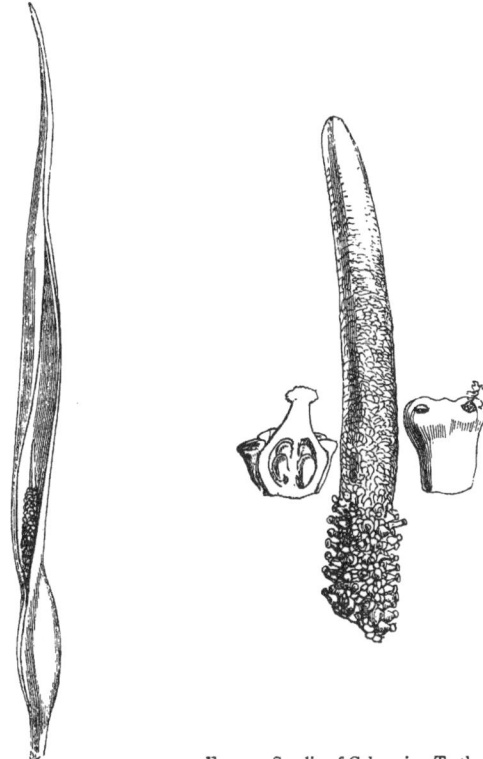

FIG. 33. Spathe with enclosed Spadix of Colocasia.

FIG. 34. Spadix of Colocasia. To the right a single subsessile anther, opening by terminal pores ; to the left a pistil cut through vertically, with surrounding scales.

stigma and the arrangement of the ovules upon the sides of the one-celled ovary in two or three series.

E

The upper belt consists of a number of stamens very densely packed, the sessile anthers cohering in sets of two, three, or more. Each stamen may be regarded as representing a single flower. Thus the structure of the flowers of Colocasia is of the very simplest kind :—

> Calyx o ; Corolla o ;
> ♂, Stamen 1 ; Pistil o ;
> ♀, Stamen o ; Pistil *syncarpous.*

2. DENDROBE (*Dendrobium nobile,* or any allied species). —At first sight neither stamen nor pistil is visible. The flower apparently consists of six delicately coloured leaves, of which five are nearly similar and broadly spreading,

FIG. 35. Flower of Dendrobe (*Dendrobium nobile*).

and the sixth, dissimilar, projects forward from the centre of the flower, with its sides curved in around the deep crimson blotch of its centre. Observe that, of these six flower-leaves, three are distinctly outside the rest, as you cannot fail to note if the flower be viewed from behind.

These are the three sepals. They are free very nearly
to the base; the two lateral ones, however, are there
united into a short, obtuse, spur-like projection at the
back of the flower. The three inner leaves of the flower
are the petals; the dissimilar concave petal with the dark
blotch being specially distinguished in this flower as the
lip. It is usual in flowers like this, in which the parts
of the calyx and corolla resemble each other in colour
and texture (as well as when a calyx only or a corolla
only is present), to speak of the envelopes of the flower
collectively as *perianth,* calling the parts of which it is
made up the *leaves* of the perianth. Before you can
ascertain the relation of the perianth to the ovary, and
whether to describe it as superior or inferior, it will be
needful to make a very careful examination of the flower.
First press down the extremity of the lip, say an inch below
its usual position. This will expose a minute conical body
coloured green, with a crimson apex projecting from the
centre of the flower, which had been previously concealed
by the incurved sides of the lip. This projecting cone it is
convenient to speak of as the *column* in flowers of this kind.
The crimson tip is the anther-cell of the solitary stamen.[1]
Remove it with the point of a penknife, and you may
observe the two waxy longitudinally grooved microscopic
pollen-masses side by side between the minute horns of the
top of the column. Observe next the under face of the
column—the face turned towards the lip. It is slightly
concave, especially immediately under the anther, where
there occurs a minute glistening depression. This depression is occupied by the stigma. The lower part of the
column is produced downwards a short way into the spur-

[1] Be careful to secure a recently expanded flower unvisited by insects,
or the anther-case and pollen may be already removed.

E 2

like projection of the lateral sepals, to which it is adnate.
From this consolidation of the stamen with the pistil the

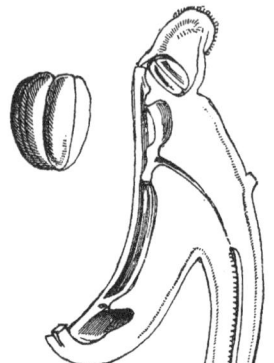

former is said to be *gynandrous.*
Lastly, make a clean vertical
section through the centre of
a flower, cutting as nearly as
may be through the median
line of the column and upper
part of the flower-stalk (*pedicel*).
No trace of the ovary will be
found in the substance of the
column, but underneath the
attachment of the perianth,
apparently in the substance of
the upper curved extremity of
the pedicel, indications are dis-
cernible of numerous rudimentary ovules, in three series, in

FIG. 36. Vertical section of the *column*
of Dendrobe, showing the pollen *in
situ.* To the left the pollen-masses
(*pollinia*) separated. The upper
part of the ovary is also shown.

the obscure cavity of the ovary. The observation is difficult in this case, because in Dendrobes the ovules are usually
not perfectly developed until a considerable period after the
time of flowering. We find, therefore, an ovary sheathed
by the adherent bases of the perianth-leaves to where they
become confluent with the pedicel.

The flower of Dendrobe may be thus described :—

> Perianth *superior, gamophyllous, irregular.*
> Stamen 1 (*monandrous*), *gynandrous.*
> Pistil *syncarpous,* ovary *inferior.*

3. CRINUM ASIATICUM (known as Sukh-dursan, Buro-
kanoor, or Tolabo).—Separate a single flower from one of
the many-flowered clusters in which they grow. The
perianth consists of a long, nearly cylindrical, or obtusely

angular tube, and a spreading limb of six equal linear
spreading or recurved segments equalling or rather shorter
than the tube. The six segments of the limb are in two
series—an inner and an outer—of three each. Lay open
the side of the perianth-tube, from top to bottom, with a

FIG. 37. *Crinum asiaticum.*

penknife, continuing the section to the base of the short
stalk of the flower. The section will exhibit the inferior
three-celled ovary adherent throughout to the base of the
perianth-tube. From the top of the ovary the long style
is continued upward through the perianth-tube. The
stamens are six in number, inserted upon the perianth

(*epiphyllous*) at the base of the limb, as in Dracæna. Crinum has therefore—

> Perianth *superior, gamophyllous.*
> Stamens *epiphyllous,· hexandrous.*
> Pistil *syncarpous,* ovary *inferior.*

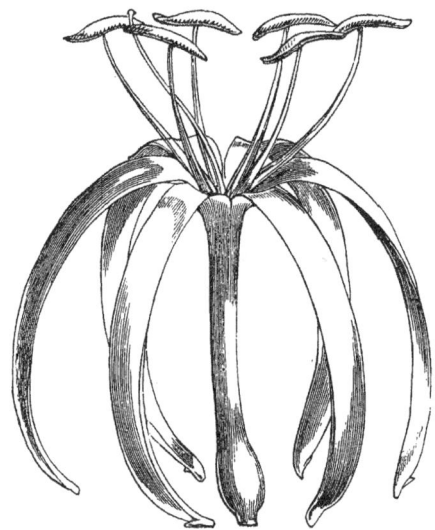

FIG. 38. Flower of Crinum.

4. DRACÆNA (any cultivated species). —The leaves of the perianth, six in number, resemble each other in size and form. The perianth is therefore regular. As the leaves of the perianth cohere by their margins to about half their length, forming a short tube, the perianth is gamophyllous. As it is wholly free from the ovary, it

may be described as inferior. There are six stamens inserted upon the perianth and opposite to its six seg-ments, there being two whorls of three each : the three outer stamens alternating with the three inner segments

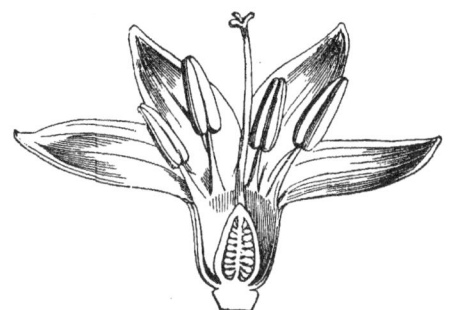

FIG. 39. Flower of Dracæna ; a vertical section.

of the perianth ; the three inner stamens alternating with the three outer stamens, consequently opposed to the three inner perianth-segments, as in like manner are the three outer stamens opposed to the three outer perianth-segments. The pistil has a three-lobed stigma and a three-celled ovary. Dracæna may be described thus :—

Perianth *inferior, gamophyllous.*
Stamens *perigynous, hexandrous.*
Pistil *syncarpous,* ovary *superior.*

5. WHEAT.—We have here an arrangement of parts widely different from that obtaining in any of the plants hitherto examined. The flowers are arranged in short, broad, sessile *spikelets,* which spikelets are disposed alternately in two rows along the top of the stem, forming a dense obtusely four-cornered *spike.* Break the entire spike in two

about the middle, and take one of the lowest spikelets from the upper half. Observe that it is attached to the stem (axis of the spike, called the *rachis*) by its *side*. In some grasses, as Rye-grass, the spikelets are attached by their *edge* to the rachis. Each spikelet consists of a pair of nearly opposite, hard, dry, scaly leaves, called the *outer glumes*, which enclose three to five closely imbricated flowers, arranged alternately on opposite sides of the axis of the spikelet. Each flower is enclosed between a *flowering-glume* and a *pale*. The flowering-glume and pale are opposite to each other, and inserted very nearly at the same point : the flowering-glume, however, is the lower, and usually embraces the pale with its incurved edges. It is similar in form and texture to the outer glumes, and often terminates in a bristle (*awn*). The pale is generally easily distinguished by its having two lateral nerves and no midrib, indicating, apparently, that it may be composed of two organs cohering together. Between the flowering-glume and the pale are the three free stamens and the superior ovary crowned with two plume-like stigmas. Note also two very minute scales, called *lodicules*, representing a perianth, inserted under the ovary. Between the outer glumes and the lowest flowering-glume of the spikelet in some grasses,

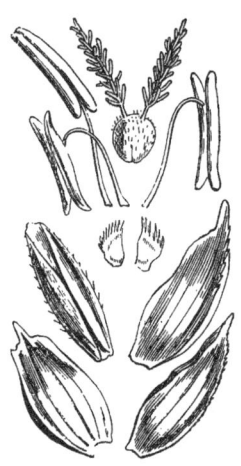

Fig. 40. The two outer glumes of a spikelet and the parts of a single floret of Wheat. The two lowest scales, right and left, are the outer glumes ; of the next pair, the scale to the right is the flowering-glume, that to the left the pale. Then come the two minute lodicules, the three stamens, and the pistil.

and, in others, above the uppermost perfect flower of the spikelet, there are one or more *empty glumes*, which are called, by some botanists, *sterile flowers*.   Occasionally a staminate flower is borne in the axil of the glume next below or above the perfect flower.   Wheat may be thus described :—

Spikelets *sessile*, with two *outer glumes*.
Flowers with one *flowering-glume*, one *pale*, two *lodicules*.
Stamens *triandrous, hypogynous*.
Pistil *syncarpous*, ovary *superior*.

6. Let us now proceed to review, as before, the five plants last examined, viz. Colocasia, Dendrobe, Crinum, Dracæna, and Wheat.

They all happen to be herbaceous plants.   The leaves, excepting in Colocasia, although narrowed below more or less, do not present an abrupt distinction of petiole and blade, and, with the same exception, the veins of the leaves are parallel and not irregularly netted.   Those which have the essential organs of the flower enclosed in a perianth have the leaves which compose it arranged in two whorls (corresponding to calyx and corolla respectively) of three each. We find our plants generally marked by (1) the absence of any abrupt distinction between blade and petiole ; (2) parallel-veined leaves ; and (3) the parts of the flowers in threes.

7. We must now soak a few grains of Wheat for comparison with the seeds of Dicotyledons. We must, however, be careful not to regard the grain of Wheat as a seed corresponding to that of the Pea or Orange, for it is a fruit, consisting of pericarp (ovary) and seed ; the pericarp being closely adherent to the true seed.   In the Crinum,

Fig. 41.   Longitudinal section of a Grain of Wheat. The embryo is represented at the base of the Seed.

Dracæna, and other plants just examined, the seeds are free from the pericarp, as they are also in Dicotyledons generally : the adhesion, in this case, may be regarded as accidental, though it is very characteristic of the fruit of grasses. Cutting the grain open, we find the embryo near the base occupying about one-fourth or one-fifth of its contents, the rest of the seed being filled with a starchy albumen.

8. The structure of this embryo we must endeavour to understand, though in order to make it clearly out very careful sections must be made through it lengthwise. The accompanying cut will supply a good idea of the arrangement of its parts. We do not find the first leaves of the embryo opposite to each other, forming a *pair* of cotyledons, as in the Bean and other Dicotyledons, but they are *alternate;* the outermost only being regarded as a seed-leaf or cotyledon. Those which it sheaths belong to the plumule. The cotyledon being single, the embryo of Wheat is called *monocotyledonous.* The lower part of the embryo is the radicle. This never directly elongates in germination, but the internal, rudimentary root-buds, *r,* burst through it and develope into the root-fibres of the plant. The process of germination is similar to that of Dicotyledons, with this difference in regard to the origin of the root : the sheathing portion of the cotyledon is protruded from the seed, and embraces the base of the plumule, which ultimately developes into a stem.

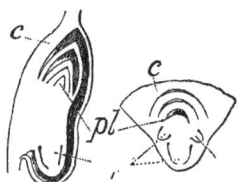

FIG. 42. Longitudinal sections, cut at right angles, of the Embryo of Wheat, showing the cotyledon *c,* the plumule *pl,* and the root-buds, *r.*

Now a structure similar to that of Wheat we find in the seeds of other corn-plants. In Rice the grain (fruit) presents a different appearance, owing to the circumstance that in this plant the fruit is closely invested by the pales, which form a sort of spurious pericarp, as in Barley and Oats of temperate countries, in which species the fruit becomes actually adherent to the investing pales, which can only be removed by grinding. The seeds of all of these, however, are albuminous, corresponding, in this respect, to the seed of the Castor-oil plant.

9. There are plants presenting so many characters in common with the five last examined as to be universally classed with them, although they may differ from them in the absence of albumen in the seed, and in other points, just as the Pea differs from the Castor-oil amongst Dicotyledons. But whether albuminous or exalbuminous they are all MONOCOTYLEDONS, with rare exceptions, so that this term comes to be applied to a second great Class, just as Dicotyledons is applied to the members of the class of which we previously examined representative types. The five plants which we have just examined all have monocotyledonous embryos, excepting Dendrobe, which is exceptional, in having an embryo destitute of a cotyledon; they all have parallel-veined leaves, excepting Colocasia; and those with a perianth have its parts in threes. Now these characters, supported by others derived from the internal structure of the stem, are, as a rule, common to all Monocotyledons, and in contrast to those characters which we have shown to prevail amongst and to mark Dicotyledons.

All flowering plants are either

DICOTYLEDONS or MONOCOTYLEDONS.

Embryo . .	With 2 cotyledons, the radicle itself usually elongating.	With 1 cotyledon, the radicle usually remaining undeveloped.
Leaves . .	Net-veined.	Straight-veined.
Perianth . .	Parts in 4's or 5's.	Parts in 3's.
Wood . .	In a continuous ring.	In isolated bundles.

11. Upon characters afforded by the flower, of subordinate importance (because less constant) to those which distinguish Dicotyledons from Monocotyledons, botanists divide each Class into Sub-classes and Divisions. The kind of characters upon which these Sub-classes and Divisions rest we are already familiar with, having examined representatives of each. The Divisions are further divided into numerous Orders. These are treated of in subsequent lessons. The Sub-classes and Divisions may be synoptically arranged thus :—

12. DICOTYLEDONS are, in respect of envelopes of the flower—

Dichlamydeous (*Dichlamydeæ*), as Poppy, Mustard, Rose, Melastoma, Zinnia, Rose Periwinkle, Basil,—or

Incomplete (*Incompletæ*). If incomplete, either

Monochlamydeous (*Monochlamydeæ*), as Grass-cloth Nettle,—or

Achlamydeous (*Achlamydeæ*), as Willow.

Dichlamydeous flowers are either

Polypetalous (*Polypetalæ*), as Poppy, Mustard, Rose, Melastoma,—or

Gamopetalous (*Gamopetalæ*), as Zinnia, Rose Periwinkle, Basil.

Polypetalous flowers have their stamens inserted on the receptacle (hypogynous), and are hence called—

Thalamifloral (*Thalamiflora*), as Poppy and Mustard,—or, inserted upon the calyx (perigynous or epigynous), and are hence called—

Calycifloral (*Calyciflora*), as Rose and Melastoma.

13. MONOCOTYLEDONS have their flowers, often imperfect, and with or without a minute, scaly perianth, arranged upon a spadix, hence called—

Spadicifloral (*Spadicifloræ*), as Colocasia,—
 or with a perianth of petal-like leaves, hence called—
Petaloid (*Petaloideæ*), as Dendrobe, Crinum, Dracæna—
 or with chaffy glumes or scale-like bracts enclosing the
 flowers, hence called—
Glumaceous (*Glumiferæ*), as Wheat.

Petaloid Monocotyledons have their perianth—
 Hypogynous (*Hypogynæ*), as Dracæna,—or
 Epigynous (*Epigynæ*), as Dendrobe and Crinum.

14. The Classes, Sub-classes, and Divisions may be
tabulated thus :—

Flowering Plants (*Phanerogamia*).	DICOTYLEDONS	Dichlamydeæ	Polypetalæ	Thalamifloræ.
				Calycifloræ.
			Gamopetalæ.	
		Incompletæ		Monochlamydeæ.
				Achlamydeæ.
	MONOCOTYLEDONS			Spadicifloræ.
			Petaloideæ	Hypogynæ.
				Epigynæ.
				Glumiferæ.

CHAPTER VI.

1. YOU may now begin to examine and describe any flowers which may be within reach. And, in order that your work may be of value, I give at page 65 a form or schedule employed (under a very slightly different form) by the late Professor Henslow of Cambridge, both in his university and village-school teaching, the purpose of which is to compel attention to those points which are of the first importance (because most constant) in the structure of flowers. A supply of these schedules should be kept on hand for daily use.

Most of the terms made use of in filling up the schedules you have already acquired. It may be well, however, to look over the following list, which embraces all that need be used at present in describing the flower in the columns of the schedule, and if any have been forgotten, to turn back to them, by referring to the Index.

2. The column headed No. (number) is to be filled with the real number of parts, whether free or coherent, in each of the four series of organs (calyx, corolla, stamens, and pistil) which compose the flower. Thus, in the Mustard-flower there are four free sepals ; and in Rose Periwinkle and Sweet Basil five coherent sepals. This number must, therefore, be entered opposite to *sepals*, under the No. column, and so on. These numbers, or a 0 opposite to an organ, necessarily indicate Suppression, when such occurs. Thus in Basil, with five sepals and five petals, there are but four stamens, one being suppressed, as we infer from the general constancy with which the parts, in each series of the flower in plants generally, correspond in number or are multiples. We often find, however, more direct evidence in the presence of a rudiment of the suppressed organ.

3. The column headed Cohesion is to be filled up with those terms which express or involve cohesion of parts, or the absence of cohesion. Thus, were the Orange-flower being described, *gamosepalous* would be entered in this column, opposite to calyx ; the calyx being gamosepalous owing to cohesion of the sepals. Poppy and Mustard, on the other hand, would be described in the same place as *polysepalous*, the calyx being polysepalous from the absence of cohesion of the sepals.

4. The last column, headed Adhesion, is for terms which, in like manner, express or involve adhesion of parts, or the absence of adhesion. Thus, in the case of the three plants just referred to, *inferior* would be entered in this column opposite to calyx, the calyx being inferior because there is no adhesion between it and the ovary. Were Melastoma

being described, the term *half-superior* would be entered in the same place, as in this plant the calyx becomes half-superior, from the partial adhesion of its tube to the ovary.

5. The terms employed in filling up schedules are :—

Of the CALYX (cohesion or its absence), *polysepalous, gamosepalous ;* (adhesion or its absence) *inferior, superior.*

Of the COROLLA (cohesion or absence of same), *polypetalous, gamopetalous, regular, irregular ;* (adhesion or its absence) *hypogynous, perigynous, epigynous.*

Of the STAMENS (cohesion or its absence); as it is important to note the number of stamens, and not simply to write *polyandrous* when the stamens are free, whatever their number may be, as you write polysepalous and polypetalous of calyx or corolla when their parts are separate, write before the termination *-androus* the Greek numeral prefix denoting the number of free stamens, thus :—

If 1,   2,   3,   4,   5,   6,   7,   8,   9,   10.   more than 10,
mon- di- tri- tetr- pent- hex- hept- oct- enne- dec- poly-androus.

If the stamens cohere by their *filaments*, they are *mon-, di-,* or *poly-adelphous ;* if by their *anthers*, they are *syngenesious.*

(Adhesion or its absence) *hypogynous, perigynous, epigynous, epipetalous, gynandrous.*

Of the PISTIL (cohesion or its absence), *apocarpous, syncarpous ;* (adhesion or its absence) *superior, inferior.* To denote the number of carpels constituting the pistil, whether they be free or coherent (if the latter, the number being inferred from the divisions of the style or stigma), the same Greek numerals as are employed to indicate the number of stamens are prefixed to the termination *-gynous.* Thus *monogynous* signifies with one style or stigma, *polygynous* with many styles, or stigmas, or distinct carpels. I have omitted these terms in the schedules of the Type-species in Part II. of this book, simply noting whether the pistil be apocarpous or syncarpous. The number of carpels is given in the No. column.

Of the PERIANTH (cohesion or its absence), *polyphyllous, gamophyllous (regular, irregular)*; (adhesion or its absence) *inferior, superior.*

6. The schedule here given, by way of example, is filled up from the flower of the Orange, with the characters of which you ought to be sufficiently familiar by this time.

Organ.	No.	Cohesion.	Adhesion.
Calyx. sepals.	5	Gamosepalous.	Inferior.
Corolla. petals.	5	Polypetalous (regular).	Hypogynous.
Stamens. filaments. anthers.	∞	Polyandrous. Polyadelphous.	Hypogynous.
Pistil. carpels. ovary.	∞	Syncarpous.	Superior.
Perianth. leaves.	†	†	†

Class.	Division.	Name.
Dicotyledon.	Thalamifloræ.	Orange.

N.B. The sign ∞ denotes *many.* No entry is made opposite to perianth (†), because it is described as calyx and corolla in the case of the Orange and other dichlamydeous Dicotyledons.

F

CHAPTER VII.

THE VARIOUS ORGANS AND THEIR MODIFICATIONS.

1. Further examination of Plant-structure. The importance of frequent practice in order to acquire facility in the use of terms.
2. (Organs of Nutrition.) The Root originates from? Tap-root; fibrous root. Adventitious roots. Roots sometimes become thickened and tuberous.
3. The Stem originates from? Axillary and terminal buds. Direction assumed by stems. Rhizome; tubers; bulbs. The Stock. Remarkable modifications of stem-structures.
4. Leaves always lateral organs. Their arrangement upon the stem. Nodes and Internodes.
5. Cotyledonary leaves are temporary. Scale-leaves. Duration of leaves.
6. Petiole and Blade. Vernation. Venation.
7. Outline of leaves. Simple and compound leaves.
8. Simple undivided leaves.
9. Simple divided leaves.
10. Compound leaves.
11. Apex and base. Mode of attachment to the stem. Margin. Surface.
12. Stipules. Stipulate and ex-stipulate.
13. Remarkable modifications of leaves. Phyllodes.
14. (Organs of Reproduction.) Arrangement of flowers upon the stem. The principal kinds of inflorescence.
15. The Bracts; bracteate, ebracteate. Involucre.
16. Æstivation of the calyx and corolla.
17. Parts of a petal. Of a gamopetalous corolla, and of a gamosepalous calyx.

1. WE now proceed to examine some of the different forms assumed by the Vegetative, or, as we previously termed them, the Nutritive, organs of Plants—viz. the Root, Stem, and Leaves. Also, so far as previous chapters leave it needful, the general character of the Reproductive Organs, and especially the structure of the Fruit.

In order to avoid ambiguity, we shall find it necessary to employ not only the substantive terms used by botanists to denote the several organs themselves, but also the more important of the adjective terms employed to denote special modifications of the same. The necessary terms are very simple, and easily learned, and, with moderate perseverance, facility in applying them may soon be acquired.

By carefully examining one plant every day, first filling up a schedule from the flower, and then writing out, with these lessons before you, a description of all the organs in detail, very considerable progress will be made in practical botany in the course of a single season.

In this and the following chapter, devoted to the structure of the various organs, whenever it has appeared desirable for the sake of illustration, I have named a common plant, which

may be referred to; but it must always be borne in mind that
the organs of plants—root, stem, leaf, and flower—are very
prone to accidental variation, especially in minor particulars,
so that occasionally I may be found apparently contra-
dicting Nature : but in such cases do not rest satisfied with
reference to a single specimen ; compare together a number
of specimens whenever it can be done, and you will then
find, I believe, the illustration confirmed.   In explaining
the terms used to denote the mere outline and form of
organs, I have not generally referred to any illustration.
I leave them to the learner to find out for himself.

2. The Root.—In the germinating Pea we find that the
root is developed by the direct downward elongation of the
radicle of the embryo.   A root thus originating forms what
is termed a *tap-root.*   We have good examples of this
primary root-axis in a large number of Dicotyledons, both
trees and herbaceous plants, though in many, by arrest, or
by repeated branching, it loses, more or less, its character
as a proper tap-root.   In Monocotyledons, owing to the
origin of the root from root-buds which burst through the
undeveloped radicle of the embryo, we never have a tap-
root.   In these plants the root generally consists of
numerous independent fibres, branched or unbranched.
It may be described simply as *fibrous.*   Pull up any grass
and you will find such a root.

Many plants which at first sight appear to be stemless
(*acaulescent*) we shall find possess a more or less creeping
stem, giving off root-fibres at the nodes.   This is a very
frequent condition, and many herbaceous plants, both
Dicotyledons and Monocotyledons, are principally multiplied,
and the area which they occupy extended, by such creeping, root-producing stems.   Roots originating in this way,
and not by direct prolongation of the radicle of the embryo,

are distinguished as *adventitious.* When adventitious roots
are given off by climbing or erect stems, as in the Pepper,
Banyan, and Screw Pines, and very many trees and climbing
or epiphytal herbs growing in hot, moist, Indian forests,
they are termed *aerial.*

FIG. 43. Screw Pine (*Pandanus*), showing aerial adventitious Roots.

Keeping the distinction between true and adventitious
roots in view, it will be clear, from what we have said of the
origin of the root-fibres in Monocotyledons, that they are
always adventitious. Whether true or adventitious, however, the function of the root is the same.

The root frequently becomes much thickened in perennial and biennial herbaceous plants, serving as a reservoir of nourishment for the growth of the sprouts of the following season. When the branches or fibres of a root become thickened in this way, the root is said to be *tuberous.* Such tuberous roots much resemble certain forms of underground and similarly thickened stems, but differ from them in the absence of leaf-buds. Potatoes and Onions are called roots, but we shall presently show that this is a misnomer.

FIG. 44. *Cycas revoluta.* The stem (*caudex*) unbranched.

3. The STEM always originates in a *bud;* the primary stem of the plant from the bud of the embryo—the plumule. Branches in like manner originate in similar buds formed in the axils of leaves. Buds borne in the axils of leaves are *axillary;* those which terminate a stem or branch, and which, after a winter's rest in cold climates, renew the shoot, are *terminal.* Some trees, as Palms and Cycads, never or rarely develope any other than a terminal leafy bud,

excepting when they form a flowering branch.   The consequence is that their stems remain unbranched.

We have already referred to the distinction between
woody and herbaceous stems.   Besides erect or ascending
annual flowering-stems, many herbaceous plants possess a
stem which either creeps upon the surface of the ground,
or which spreads wholly under the surface, giving off leafy
and flowering shoots above and roots below.   This underground form of creeping stem is called a *rhizome.*   Beginners
are very liable to regard it as a root, and such is the
common notion respecting it.   The capacity of developing
leaf-buds at regular intervals, and the presence of leaves in
the very reduced form of minute scales, indicate its true
stem character.   In many plants, spreading underground
stem-branches become greatly thickened, like tuberous roots,
and serve the same end in the economy of the plant.   We
have a good example in the Potato and Yam.   The "eyes"
of the Potato are leaf-buds, and shoots develope from them
when planted or placed in damp cellars.   Such thickened
portions of underground stem are called *tubers.*   Somewhat similar is the very short and abruptly-thickened base
of the erect stem of some herbaceous plants distinguished
under the name of *corm.*   In the Onion, Lilies, and Crinum
we have an analogous condition, disguised by very numerous, much thickened, scale-like leaves.   If we peel off
these scales successively, until they be all removed, we find
a flattened or conical solid base remains, from the under
side of which root-fibres are given off.   This portion is the
excessively shortened stem from which the tall flowering
peduncles arise.   Stems of this kind, with the internodes
suppressed and covered by thickened, scaly leaves, are
called *bulbs.*   They may be regarded as equally leaf and
stem formations.

Perennial herbs, the flowering and leafy stems of which die down annually, often form a permanent tufted mass, called a *stock,* either wholly or partially hidden under the surface of the ground. The stock results from the persisting bases of the leafy stems. From the axils of the scale-like leaves which these persisting stem-bases bear the annual shoots are thrown up each spring. The passage from plants with this form of perennial stock to those in which more of the exposed portion of the stem is perennial, as in bushes, shrubs, and trees, is quite gradual.

The form of stems and the direction which they assume above ground are exceedingly varied. Most of the modifications which they present are denoted by terms in ordinary use. Thus the stem may be erect, procumbent, or prostrate; cylindrical, angular, furrowed; and so on.

Branches sometimes assume very anomalous forms, and might be mistaken for distinct structures, as in the *spines* of some varieties of Orange and Lime, or the common Flacourtia sepiaria, and in the *tendrils* of the Grape-vine and many garden Trumpet-flowers (*Bignonias*). All spines and tendrils, however, are not arrested or specially modified branches; they are often leaves or leaf-appendages.

The internal structure of the stem may be more suitably described when we speak of cells and tissues.

4. LEAVES.—We have already spoken of leaves as originating around the growing apex of the stem as minute, cellular projections. They are never terminal organs; neither are they, normally, capable of forming buds upon their surface. The arrangement of the foliage-leaves upon the stem, though at first sight it may appear accidental, is according to a law generally constant in the same kind of plant. Compare, with respect to leaf-arrangement, a young shoot of Tamarind with one of the Orange or Banyan. Try to find two leaves

exactly or nearly in the same vertical line, one above the other. In the Tamarind the upper leaf will be removed from the lower by but two internodes; in the Orange and Banyan, by either five or eight. When a single leaf is given off at each node, the leaves are said to be *alternate;* if a pair of opposite leaves, they are described as *opposite;* if three or more in a whorl, as *verticillate.*

The general arrangement of the leaves is materially affected by the extent to which the internodes of the stem develope. In the Chinese Primrose of gardens, and Ranunculus, we find the lower leaves springing in a tuft from the short stock, owing to the non-development of the lower internodes; while in the latter the upper leaves are separated from each other by distinct, and often long, internodes. A parallel but more remarkable case we see in American Aloe (*Agave*) and Adam's Needle (*Yucca*), much cultivated in India, in which plants a succession of (really alternate) leaves are given off from a very short stem or stock, the internodes of which are not perceptibly developed. This is continued until the approach of the flowering season, when the stem suddenly begins to lengthen out, and the leaves gradually decrease in size to mere scales. In Crinum and Tacca they cease altogether for a long interval, leaving the flower-stalk naked. In most deciduous trees, the internodes from which foliage-leaves are given off are tolerably uniformly developed, but in Pine and Deodar an anomalous condition occurs. In these trees there are two kinds of leaf—one a small, membranous, brown scale-leaf: the other, a green, needle-like leaf. The needle-like leaves are arranged in tufts of 2, 3, or 5 in Pines, or in many-leaved clusters in the Deodar, in consequence of the non-development of the internodes of the excessively short branches which bear them. That these tufts really arise from shortened branches is obvious on examination, for they occupy the axils of the smaller scale-leaves, and some of them occasionally develope their internodes, when, conquently, the needle-leaves are borne singly upon the shoot, and are separated from each other by more or less marked internodes.

5. The cotyledons are the first leaves of the primary axis of the plant. They are usually, but not always, very shortlived, and shrivel up and die at an early stage. In some plants they never leave the testa of the seed, but remain underground, while in others they rise above the surface and assume partially the functions of ordinary foliage-leaves. The first leaves of branches ordinarily differ from those which follow, in being much smaller, and often, in certain Natural Orders and in cool climates, in being hard and scaly. These are the *scale-leaves*. They serve as protective organs to the delicate rudiments of the foliage-leaves which they enclose, and into which they usually pass more or less gradually, thus convincingly showing that they are both modifications of one and the same organ.

Many trees develope each season terminal as well as axillary buds. As before pointed out, it is only by the development of the former that the original stem or its branches are prolonged. Some plants never renew their branches by annual terminal buds, while others annually develope branches from both terminal and axillary buds. This variety of conditions in respect to the relations of terminal and axillary buds has much to do with the general aspect of the tree.

Leaves vary in their duration. They may last but one season, at the close either separating by an articulation from the stem, leaving a clean scar, or remaining attached and gradually decaying. In Evergreens, the leaves last two or more seasons; in some Pines, indeed, they persist for several years.

6. In the fully developed leaf we have already distinguished petiole and blade. The mode in which the blade is folded while enclosed in the bud is spoken of as the *vernation* of the leaf.

The blade is divided into symmetrical halves by a *midrib*, which, continuous with the petiole, runs from the base of the blade to its apex. In the Begonias, several of which are wild, and others cultivated in India for the sake of their beautiful, variegated leaves, the sides of the leaf are more or less oblique or unequal.

The arrangement of the veins in the blade is made a special study by botanists who concern themselves with fossil plants, for the *venation* of leaves is almost the sole character left them of importance in fossil impressions, upon which to base comparisons with species still living.

7. In describing plants, the form or outline of the leaf must be noted, and an appropriate adjective term selected to express it. As the forms assumed by leaves are infinitely varied, it necessarily follows that numerous terms must be used to denote them. The same terms apply, however, to any organs with plane surfaces, whether foliage or flower-leaves. The more important only we can note here.

In the first place, compare the leaf of a Mango or Banyan with one from the Rose, Litchi, or *Sterculia fœtida*. You observe that in the two former the leaf is in one piece; in the three latter the petiole bears several distinct pieces. These distinct pieces are called *leaflets*, and leaves which are thus divided into distinct leaflets are termed *compound*. Leaves, on the other hand, which are not divided into separate leaflets are termed *simple*. Simple leaves are frequently deeply divided, but the divisions do not extend to the base

of the blade, nor are they separately jointed to the petiole.
The portions of a simple leaf thus divided are called the
segments or *lobes* of the leaf.

8. *Simple undivided leaves.*—It will be useful practice to
try to find leaves which correspond to the various outlines
figured below.   It will constantly happen that the form of
some leaves may be as correctly described by one term as

FIG. 45. Simple Leaf.   FIG. 46. Compound (pinnate) Leaf.

by another; and again, some leaves vary so much on the
same branch that they may be found to match two, three,
or more of the outline figures.   In describing such leaves
you must use the terms which denote the usual extremes of
variation, as "leaves varying from lanceolate to ovate," or
from "oblong to elliptical," &c.   The terms may also be
combined when needful, as *oval-oblong, linear-lanceolate.*

9. *Simple divided leaves.*—We may class these under two
series—viz. (1) those in which the segments radiate from
the extremity of a petiole, and (2) those in which they are

given off successively from a midrib. The former are of the *palmatifid*, the latter of the *pinnatifid* type. If the segments be separated nearly to the petiole, the leaf is *palmatipartite ;* if nearly to the midrib it is *pinnatipartite ;* the termination *-partite* being substituted for *-fid* to denote deep division of

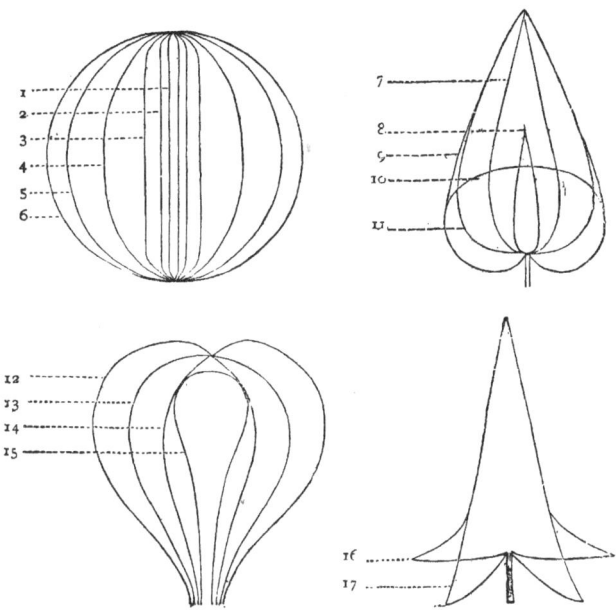

FIGS. 47, 48, 49, 50. Outlines of Simple Leaves.

1. Acicular.	7. Lanceolate.	13. Obovate.
2. Linear.	8. Subulate.	14. Oblanceolate.
3. Oblong.	9. Cordate.	15. Spathulate.
4. Elliptical.	10. Reniform.	16. Hastate.
5. Rotundate.	11. Ovate.	17. Sagittate.
6. Orbicular.	12. Obcordate.	

the blade. There are many modifications of these principal types of form, distinguished by special terms, but with these it is not needful to burden the memory just at present.

10. *Compound leaves* we shall class as we have done the divided simple leaves, from which they differ in having the blade divided into *leaflets*, which are given off from the summit of the petiole, or from the midrib (common petiole), as the case may be. The leaflets separate from the petiole or midrib in the same way that the entire leaf separates from the stem, that is, without tearing. Sometimes it is difficult to tell whether a leaf should be called simple

FIG. 51. Pinnate Leaf. A pair of adherent (*adnate*) stipules are shown at the base ; one on each side.

or compound. Generally, however, it is plain enough. Many beginners fall into the mistake of calling *leaflets* leaves, but a little care will rarely fail to save any one from such a mistake. Compound leaves are either of the *pinnate* type, as Rose and Tamarind, or of the *digitate* type, as *Sterculia fœtida*. The Rose leaf is *unequally pinnate*, because there is an odd leaflet at the end of the common petiole. When the odd leaflet is absent, as in Tamarind, the leaf is *equally* or *abruptly pinnate*. A leaf becomes twice pinnate (*bipinnate*) when the common petiole, instead of bearing leaflets, bears secondary petioles upon which the leaflets are pinnately arranged.

When leaflets are arranged on the digitate plan, and are but three in number, they are called *ternate*, and the leaf is *tri-foliolate;* if five, *quinate*, the leaf being *quinque-foliolate.* The leaflets may be twice ternate (*bi-ternate*) if the petiole bears three secondary petioles, each of which bears three leaflets, and so on.

11. The point of a leaf or leaflet at which the midrib

FIG. 52. Digitate Leaf.

ends is called the *apex.* The point where it passes into the petiole, or, if the leaf be sessile, where it is joined to the stem, the *base.* The *apex* and *base* vary considerably in outline, and attention must be paid to both in describing the form of leaves. The *apex*, if sharp, is termed *acute;* if blunt or rounded, *obtuse;* if with a very shallow notch, *emarginate;* if the notch be deep, *bifid*—the leaf becoming

bipartite if divided nearly to the base; it is *trifid* or *tripartite*
if there be three divisions. The base of the blade in cordate, sagittate, and hastate leaves we have already figured.
If the base of a sessile leaf clasp the stem, it is termed
amplexicaul. If the lobes at each side of the base of an
amplexicaul leaf be united together on the side of the stem
opposite to the midrib, so that the stem appears to pass
through the blade, the leaf is *perfoliate.* If the bases of
two opposite leaves be united on each side of the stem,
the leaves are said to be *connate.* Sometimes, in sessile
leaves, the margins of the blade are continued down the
sides of the stem, forming wings to it. Such leaves are
decurrent. When the petiole joins the blade upon its under
surface and not at the margin, as in the Sacred Lotus
(*Nelumbium*), the leaf is said to be *peltate* (Fig. 54).

The *margin* of the leaf varies, being sometimes perfectly
continuous and not indented or toothed at all, when it is
termed *entire;* it is *serrate* if with sharp teeth directed forward, like those of a saw ; *dentate* if with sharp teeth directed
outward ; *crenate* if with rounded teeth.

The *surface* may be more or less hairy, or altogether
without hairs, when it is termed *glabrous.* Different terms
are used to denote different kinds and degrees of hairiness,
but it is not important to learn these at present.

12. Taking up again a specimen of the Pea or Rose,
observe on each side of the base of the petiole a leafy organ
somewhat resembling a leaflet. In the Pea these organs
are very large—larger, indeed, than the leaflets. These are
the *stipules.* Leaves provided with stipules are called
stipulate, and leaves destitute of them, as those of Mustard,
exstipulate. Like leaves and leaflets, the stipules vary in
form, but they are usually small, and often fall away very
early, as in the Banyan and Bread-fruit.

13. Foliage-leaves are sometimes curiously modified, either to serve some special purpose, or by the absence of the blade, or the reduction of the leaf to a mere spine. Thus in the Pea we find the extremity of the common petiole and two or more of the lateral leaflets assume the form of *tendrils*, enabling the weak stem to lay hold of

Fig. 53. Pinnate (quadrifoliolate) leaf, with stipules. *Stip.* stipules.

supports in climbing. Compare with the tendrils of the Pea those of the Grape-vine, which we have described as branches modified for the same purpose (p. 72).

In Barberry and its allies, common in the Himalayas, the first leaves borne by the branches are reduced to sharp

spines, from the axils of which spring tufts of ordinary
foliage-leaves, borne (as in Pine and Cedar) upon branches
with undeveloped internodes.   Leaves tufted in this way are
said to be *fasciculate.*   Stipules also are sometimes replaced

by spines. When the blade of the
leaf is absent, the petiole sometimes becomes flattened to such
an extent as to look like an
entire leaf, in order to replace
the blade as an organ useful to
the plant.   But the flattening
is generally vertical, so that the
apparent leaf is placed edgewise
upon the stem, instead of spreading horizontally.   By this character these leaf-like petioles may
be generally recognised.   They
are called *phyllodes.* Sometimes
the true blade is partially developed at the extremity of the
phyllode, thus putting its petiolar
character beyond doubt.

14. We cannot fail to have
observed the various ways in
which the flowers are borne
upon the stem, in gathering and
comparing together the common plants which we have had

FIG. 54. Sacred Lotus
(*Nelumbium speciosum*).

occasion to use in the course of these lessons.   It is
convenient to speak of the Flowering System, or mode
of arrangement of the flowers of plants, as the INFLOR-
ESCENCE.

In the Sacred Lotus we find a solitary terminal flower,

borne by a firm herbaceous peduncle, which appears to spring directly from the root. Such radical peduncles, whether they bear one flower, or many (as in Crinum and Tacca), are called *scapes.*

FIG. 55. Indian Mustard (*Brassica juncea*). The inflorescence a *raceme.*

In Mustard, the peduncle, instead of ending in a solitary flower, gives off successively a number of shortly-stalked (pedicellate) flowers in succession, until it exhausts itself and ceases to lengthen. Such an inflorescence is termed a *raceme.*

Common Plantain (*Plantago*) of waste ground has a

G 2

similar kind of inflorescence, but the flowers are sessile. This difference distinguishes the *spike* from the raceme.

The *corymb* is a form of raceme in which the lower pedicels are much longer than the upper ones.

FIG. 56. Crinum. Flowers in simple umbels borne on stout succulent scapes. Bracts spathaceous.

In Coriander, Fennel or Carrot, and Crinum, the flowers are borne upon pedicels springing apparently from one point. Such an arrangement of pedicellate flowers constitutes the *umbel*. But as you find each of the umbels in the three first-named plants borne upon peduncles, which, like the pedicels, also spring from one point, their entire inflorescence forms a *compound umbel;* the umbels of single flowers being the *partial umbels.*

Observe the ring of small leaves at the base of the pedicels in the Carrot, forming an *involucre.* In compound

FIG. 57. Vertical section through a flower-head of Zinnia. The receptacle convex.

umbels we frequently have both *general* and *partial* involucres; the former surrounding the compound umbel, the latter each partial umbel.

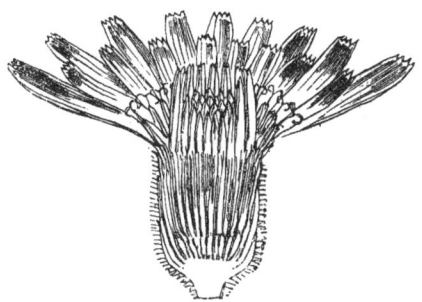

FIG. 58. Section of flower-head of Sonchus. The receptacle plane.

Suppose, now, all the flowers of a simple umbel to be *sessile.* We should have the same form of inflorescence as we find in Zinnia and Sonchus, in which a number of florets are arranged upon a conical or flattened disk (the *common receptacle*) surrounded by an involucre.   Such an inflorescence may be called a *flower-head.*   The older botanists used to regard the flower-head as a kind of compound flower, enclosed in a common calyx, but we found in the Zinnia that it was composed of a number of distinct flowers (florets), each with its own calyx and corolla.   The ring of bract-leaves which surrounds the flower-head answers to the ring surrounding the umbel, and is called by the same name—*involucre.*

In the Poppy and Sacred Lotus the peduncle (scape) terminates in a solitary flower.   In Mustard we found that the peduncle does not itself terminate in a flower, but gives off a succession of secondary branches (pedicels), each of which bears a flower.   If we take a Ranunculus, we shall find that the main or primary stem of the plant directly terminates in a flower like that of Poppy, and if, as is usual, there is more than one flower upon the plant, the 2d, 3d, 4th flowers, and so on, terminate respectively as many successive independent branches, springing from the axils of the leaves.   Such forms of inflorescence, in which the peduncle, or axis, itself terminates in a flower, are termed *definite* or *cymose,* while those inflorescences in which the principal axis never actually terminates in a flower, but, as in Mustard, gives off a succession of lateral pedicels, are termed *indefinite.*   In the St. John's Worts (*Hypericum*) and Exacum we have the cymose or definite inflorescence well shown in their characteristic, forked *cymes.*

An inflorescence which branches irregularly, like that of Melia, Litchi, Mango, and most Grasses, is called a *panicle.*

In describing the form of an inflorescence, when it does not exactly coincide with any of the principal types here defined, that which is nearest may, for the present, be applied to it in an adjective form, as *spicate, racemose, paniculate*—like a spike, a raceme, a panicle, and so on.

15. As we progress from below upwards in the examination of the various organs of the plant, we notice, in approaching the flowers, that the foliage-leaves usually decrease in size, so that those next to the flower, or from

FIG. 59. Melia. The inflorescence a terminal panicle.

the axils of which the flowers spring, are often very narrow and sometimes scale-like. Such reduced leaves, bearing flowers in their axils, are distinguished as *bracts*, and flowers springing from the axils of bracts are *bracteate*. The passage, in size, form, and texture, from foliage-leaves to bracts may

be very gradual, but in most plants the transition is rather abrupt. A ring or series of numerous bracts, enclosing flowers or pedicels, as in Zinnia and Carrot, we have called an *involucre.* In Mustard the bracts are undeveloped : hence the flowers are *ebracteate.*

16. The organs of the flower and their principal modifications we ought now to be tolerably familiar with from schedule practice. There are, however, a few characters of importance which require further attention, ap-

FIG. 60. Dhak (*Batea frondosa*).

plying particularly to the manner in which the parts of the calyx and corolla are folded while in bud (termed *æstivation*), to the form of the corolla and the structure of the pistil.

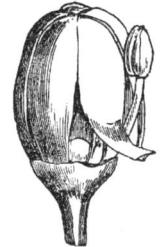

In the bud, the sepals and petals (or the lobes of a gamosepalous calyx, or of a gamopetalous corolla) may be folded with their margins either more or less overlapping, or simply in contact without overlapping. In the former case, the æstivation is *imbricate,* as in the corolla of Rose, Dhak (*Butea*), and Bignonia ; in the latter *valvate,* as in the calyx of Clematis and the

FIG. 61. Flower of Grape-vine. The petals caducous and valvate in æstivation.

corolla of Vine. Sometimes the calyx may be valvate and the corolla imbricate, as in Hibiscus.

17. The petals of a polypetalous corolla, if narrowed to
the base like those of the Mustard or Pink, are *clawed;* the
narrow part being the *claw,* the expanded part the *lamina.*
In a gamopetalous corolla, or gamosepalous calyx, the lower
united portion is called the *tube;* the free divisions, which
indicate the number of parts cohering, the *limb;* the divisions
of the limb being spoken of simply as *teeth* if small, or *lobes*
if larger.   The more important forms of the corolla are
noticed in Part II. under the groups of plants which are
respectively characterised by peculiar modifications of it.

18. The more important of the characters afforded by the
STAMENS, due to varying conditions as to cohesion, adhesion,

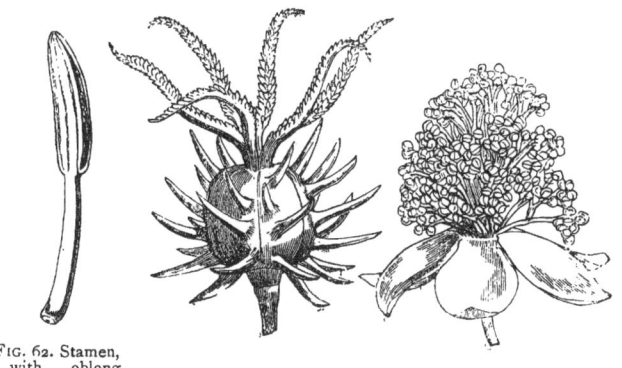

FIG. 62. Stamen,
 with   oblong
 two-celled an-   FIG. 63. Staminate and pistillate Flowers of Ricinus ; the
 ther dehiscing   stamens polyadelphous.
 longitudinally.

and suppression, have been already, directly or indirectly,
referred to.   Some other peculiar modifications which they
assume, either in form or in the mode of dehiscence of
their anthers, are pointed out in Part II. as occasion arises.

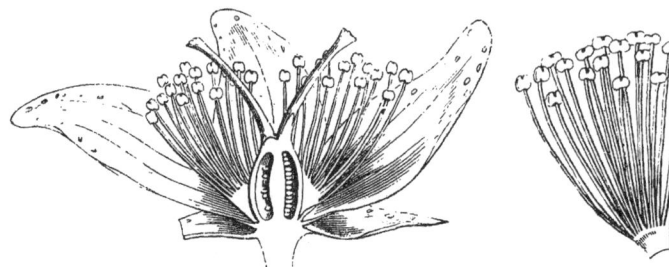

FIG. 64. Flower of Hypericum. The stamens hypogynous and triadelphous.

FIG. 65. A single phalange of Stamens of the same.

FIG. 66. Vertical section of Flower of Hibiscus. The stamens monadelphous; the anthers one-celled.

19. THE PISTIL.—When in our first chapter we spoke of
all the organs borne by the stem as leaves of some kind, you
were not in so favourable a position, as from subsequent
experience you must now be, to appreciate the broad sense
in which the word *leaf* was employed. I repeat, all the
organs borne by the stem and its branches are modifications
of one leaf-type. By this statement you are not to understand that a petal, or a stamen, or a carpel, is a modified
foliage-leaf, any more than that a foliage-leaf is any one of
these organs modified ; but they are all alike modifications
of one common leaf-type which has only an ideal existence.
Thus the leaf may be an organ either for the purpose of
nutrition or of reproduction, or it may be merely a protective organ ; but whatever function it is designed to fulfil—
in other words, whatever special organ it becomes—it is
modified appropriately to the function which it has to perform.
Thus we have the nutritive leaves, broad, green expansions,
exposing the fluids of the plant to the influence of light ;
the protective leaves, hard and scale-like, as the scale-leaves
of leaf-buds; or more delicate, and often showy and coloured,
as the enveloping leaves of the flower.

The essential reproductive leaves invariably assume one
of two forms—either that of the staminal leaf, with the blade
(the anther) thickened and its tissue partially transformed
into pollen, or that of the carpellary leaf, which is hollow,
bearing a seed-bud or seed-buds (ovules) upon its margin
inside, and terminating above in a stigma.

That this is a correct view to take of the nature of the
leafy organs of a plant we have incontestable external
evidence to prove. And this evidence is principally of
two kinds. Either we find one form of leaf passing by
insensible gradations into another, as foliage-leaves into
sepals, sepals into petals, petals into stamens, or we find

some of the leaf-organs, especially those ot reproduction, under certain conditions, assuming the character
of other organs.   Thus stamens, in many plants, have a
strong tendency to lose their character as staminal-leaves
and to assume that of petals, as you may find if you
compare a double with a single Rose.   There is, indeed, a
Rose in which *all* the organs of the flower, excepting the
sepals, so far depart from their normal character as to
become small foliage-leaves, all coloured green, and firm in
texture.

The chief difficulty in the way of accepting the notion of
the essential oneness and homology of all the leaf-organs of
a plant rests principally in the wide dissimilarity existing, in
the usual condition of things, between the leaves of the stem
and the stamens or carpels, especially the latter.   But the
acceptance and thorough appreciation of this view you will
find furnishes an invaluable key to the comprehension of all
the various modifications which the pistil and its parts, the
carpels undergo ; and it is especially with reference to these
that we shall at present concern ourselves.

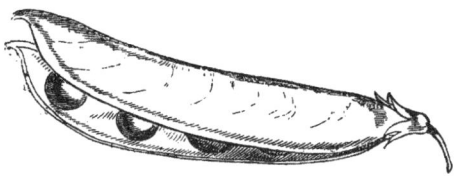

FIG. 67.   Pod (*legume*) of Pea, partially laid open to show the attachment
of the Seeds to the ventral suture.

20. Take a pistil of the simplest possible structure,—the
pistil of the Pea, Dhak, or any of their allies, for example.
You have here an apocarpous pistil, consisting of a single

carpel. Ranunculus, Nelumbium, Guatteria, and Unona also have apocarpous pistils, consisting, however, not of a single carpel, but of numerous distinct carpels.

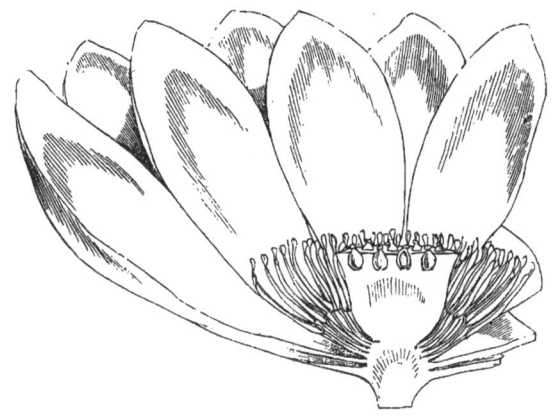

Fig. 68. Flower of Sacred Lotus (*Nelumbium*), in vertical section, showing the carpels separately immersed in the top-shaped receptacle.

A comparison of any one of the carpels of Ranunculus with the pistil of the Pea will afford satisfactory evidence that in the latter you have but a solitary carpel. In Ranunculus you observe that the stigmas are all oblique to the carpels which bear them, and that they all radiate, as it were, from the centre of the flower. A like obliquity may be noticed in the Pea, the single carpel which it possesses being the only one developed of a whorl of five. Sometimes one or more of the carpels suppressed in the Pea are

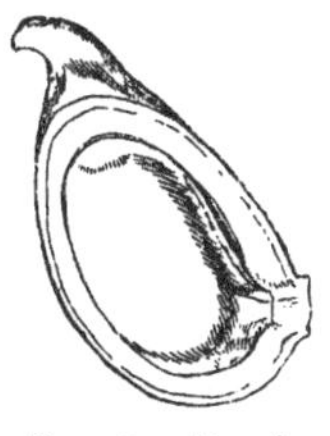

Fig. 69. Longitudinal section of fruit-carpel of Ranunculus, showing obliquity of the stigma.

developed in other species which are allied to it in general structure.

But garden Larkspur, Monkshood, and Star Anise
furnish good connecting links between
the Ranunculus and Pea ; for in these
plants you find the carpels larger than
in Ranunculus, but fewer in number,
varying from one to fifteen, and standing in a whorl around the centre of
the flower.   Each carpel of the pistil
of either of these three plants answers
to the pistil, consisting of one carpel,

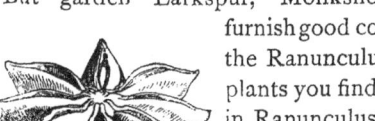

Fig. 70. Star Anise (*Illicium*).
The fruit apocarpous.   Carpels uniseriate, dehiscing by
their ventral sutures.

of the Pea.   In all of these plants the pistil is apocarpous ; the carpels, however, differ in number, as well as
in the number of ovules which they contain, and in their
mode of opening when ripe (*dehiscence*) to allow the seeds
to escape.

Observe that in Larkspur, Monkshood, and Star Anise
the ovules and seeds are borne upon the inner angle of the
carpels ; the same in the Pea ; and the inner angle of the
carpels coincides with the axis of the flower.

Now ovules are, as a rule, marginal buds (the nature and
relation of which to ordinary leaf-buds is not yet well
understood), that is, they are borne upon margins of carpellary leaves ; so we may conclude that the inner angle
of each carpel, upon which the seeds are arranged, answers
to the line of union of its infolded edges.   This line is called
the *ventral suture*.

To take the Pea again as the simplest case—if you split
it carefully open up the edge bearing the seeds, you will
find, when laid open, that half of the seeds are on one edge,
half on the other ; each margin being alternately seed-bear-
ing.   Up the middle of the opened carpel you have a strong
line or nerve (the outer angle when the carpel was closed),
which is, simply, the *midrib* of the carpellary leaf, answering

to the midrib which we find in foliage-leaves.   This line is called the *dorsal suture.*

The apex of the carpel is continued into the short style, and terminates in the stigma, which withers before the Pea is ripe.   Each of the carpels in the other plants which we have just examined presents the same features as the Pea. Ranunculus differs only in the small size of the carpels, each adapted to contain one small seed.

Suppose, now, the three or five carpels of the pistil of Monkshood, instead of being free from each other, had been developed cohering to each other by their inner faces.   The consequence would have been that we should have had a

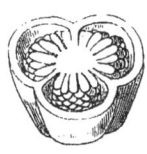   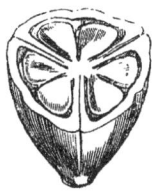   

Fig. 71. Transverse section of a three-celled Ovary, with axile placentation and multiseriate indefinite ovules.

Fig. 72. Transverse section of a three-celled Ovary, with axile placentation.   The ovules biseriate.

Fig. 73. Transverse section of a two-celled Ovary, with axile placentation & indefinite ovules.

syncarpous pistil with a five-celled ovary.   And syncarpous pistils with five cells, or more than five cells (as Orange), or fewer than five (as Lily), occur on every hand, and are nearly always explicable in this way; that is, by the cohesion of as many carpels as there are cells in the syncarpous ovary.   It follows, from this explanation of the structure of a syncarpous ovary, that each of the divisions, called *dissepiments*, by which syncarpous ovaries are separeted into distinct cells, must be double.   They must each necessarily consist of the two infolded and cohering sides of adjacent carpels. And so we often find, that when syncarpous pistils are ripe their carpels separate from each other, each dissepiment splitting into two plates.

21. From the circumstance that the ovules develope upon the margins of carpels, it must follow that when two or more carpels cohere, and their margins are infolded so as to meet in the centre of the pistil, the ovules must also be attached in the centre or axis. Their attachment, or *placentation*, as it is termed, is then termed *axile.* You find this well shown in Lily or Crinum, where there are three carpels; Solanum, where there are two; and Orange, where there are many, cohering.

But in many syncarpous pistils, although the carpels cohere, their margins are not infolded to such a degree as to reach the centre and become united there into an ovule-bearing axis. When such is the case, the placentation

is described as *parietal.* We find all grades of development of these dissepiments, from the Violet and its allies, with parietal placentation (the carpels not being infolded at all, and the ovules arranged in lines upon the

FIG. 74. Transverse section of a one-celled ovary with three parietal placentas and indefinite ovules.

inside of the one-celled ovary), to the Lily, in which the carpellary margins are inflected to the centre, and the placentation is consequently axile. The pistil of Poppy is intermediate; the margins of the numerous united carpels which compose it projecting into the cavity of the ovary without quite reaching to the centre. The placentation of this plant is exceptional, the ovules being spread over the sides of the partial dissepiments, instead of being confined to their inner edges.

In Pinks and Stitchworts the placentation is axile, but the dissepiments are lost before the ovary is fully grown, so that the ovules are collected in a head in the centre of a one-celled ovary.   Such placentation is termed *free central.*

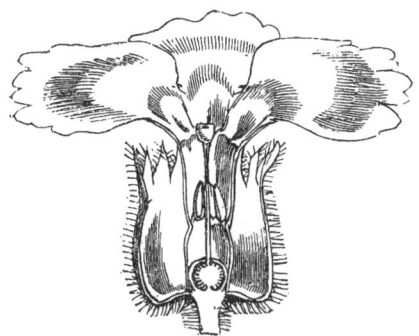

FIG. 75. Transverse section of a one-celled ovary with free central placentation.

FIG. 76. Chinese Primrose (*Primula sinensis*).

The same kind of placentation is found in Chinese Primrose, Ardisia, and Utricularia, but in these plants there is no trace of dissepiments at any stage of development of the ovary.

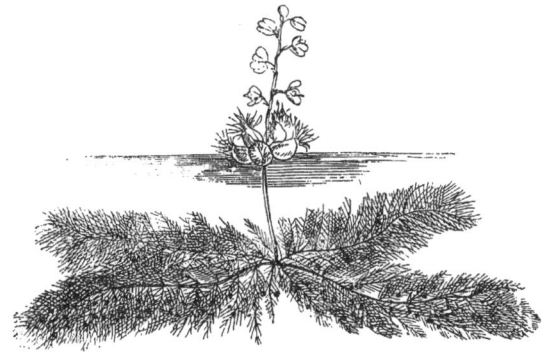

FIG. 77. *Utricularia stellaris.*

H

In Pea, Dhak, Star Anise, Unona, and other apocarpous pistils, the carpels of which contain several ovules inserted upon the ventral suture, the placentation may be described as *sutural.* The sutural placentation of apocarpous pistils is, of course, essentially the same as the axile placentation of syncarpous pistils.

22. The structure of the FRUIT deserves careful attention, especially as there is scarcely any part of the plant more liable to be misunderstood. We must learn from what part or parts of the flower the fruit results, and how to distinguish fruit from seed, for some common fruits are constantly misnamed seeds, and sometimes seeds are mistaken for fruits.

Seeds are almost invariably contained in a seed-vessel called the *pericarp,* and the pericarp may consist either of the ripened ovary only, or, if the ovary be inferior, of the calyx-tube combined with the ovary.

In the case of Ranunculus, and most plants with apocarpous pistils, the fruit consists of as many distinct carpels as there were carpels in the pistil of the flower. Each carpel contains one or more ovules in flower, and one or more seeds in fruit. The pistil we call apocarpous, and the same term applies to the fruit. In like manner we may apply the term syncarpous to all fruits which result from syncarpous pistils.

23. Now the changes which take place during the ripening of the fruit are very simple indeed in Ranunculus compared with those which take place in many other plants. We find often that an ovary with several cells in the flower is but one-celled in fruit, and that many ovules present in the flower are sometimes sacrificed during the perfecting of a single seed. Take the acorn of any Oak for example. When ripe it contains but one cell and one seed; but if the ovary be cut

across a month after flowering, it will be found to be three-celled, with a pair of ovules in each cell. This suppression of parts during the ripening of the fruit is very common. It is, however, but one of the important changes to which it is subject.

 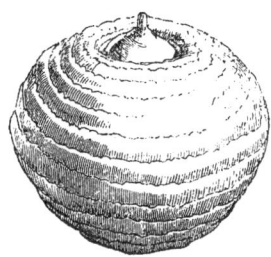

FIGS. 78, 79. Acorns of two Indian species of Oak (*Quercus*).

Another change which greatly disguises the true nature of the fruit is the excessive enlargement of certain parts, or the alteration in texture and firmness of the layers of the pericarp.

Examine a ripe Mango. You find it borne upon a peduncle. At the top of the peduncle there still remains a scar, showing where the stamens were attached, and that the calyx was inferior. A little obliquely-placed dot on the top of the Mango denotes the position of the style. It must follow, then, that the Mango-fruit has developed from the ovary only. You cut into the pulp of the fruit, and find that it encloses a hard stone. Break the stone, and the seed will be found inside. The stone is a hardened inner layer of the pericarp, the pulp a succulent outer layer; both the stone and the pulp which surrounds it originate from the walls of the ovary. Stone-fruits, like the Mango and Jujube, are called *drupes*. Like the small fruit-carpels of Clematis and

Ranunculus, they do not split open (*dehisce*) when ripe to let
the seed escap, but simply fall to the ground, where the
pericarp rots away, and the seed begins to germinate.

Try now a Loquat, Jambolan, or Guava.   Examine first
the top of the peduncle, underneath the fruit.   There is no
scar to be found as in the Mango, but at the top of the fruit
you find the distinct remains of the limb of the calyx, and
sometimes even a few withered stamens.   You thus know
the fruit to be inferior.   In each of these fruits the calyx-
tube is adnate to the ovary, so that ovary and calyx-tube
together constitute the pericarp.

Next try an Orange.   At the bottom of the fruit you find
either the calyx still remaining, or its scar.   The fruit is a
true *berry*, and the same name you may apply to any syncarpous fruit that is succulent, and that does not open
(succulent fruits rarely do) to allow the seeds to fall
out, such as the Grape and Cape Gooseberry (*Physalis
Peruviana,*).   On the top of the fruit is a little round
scar, left by the style, which soon withers after flowering,
and usually breaks off.   The Orange, therefore, is clearly a
superior fruit, developed solely from the ovary of the flower.
Cut it across, and you find it divided into a variable number
of cells by membranous dissepiments, each cell answering
to a carpel.   In the pulp which fills the cells, and which is
developed from the inside of the outer wall of the ovary, the
seeds are embedded.

Try a Cocoa-nut.   From the extraordinary enlargement
and change in texture undergone during maturation of the
fruit, probably all external indication of the position of the
perianth is obliterated.   In such cases it can only be determined by watching the development of the fruit whether
to call it superior or inferior.   In Cocoa-nut it is superior,
resulting from the ovary alone, two of the three cells of which

are early obliterated, or, rather, rudimentary from their origin, so that the fruit is one-celled and one-seeded : the triangular form of the fruit still indicates its tricarpellary character. A tranverse section through the entire fruit shows a thick outer layer of the pericarp, fibrous in texture, and a thin bony inner layer (the shell). The cavity of this inner layer (*endocarp*) is occupied by the seed. The seed is hollow, consisting of a uniform layer of solid albumen closely applied over the inner surface of the endocarp, with a portion (the milk) unconsolidated in the cavity, and a minute embryo occupies a little cell in the albumen at the base of the nut.

A syncarpous fruit that is dry when ripe, and which opens (*dehisces*)—either by the pericarp splitting from the top to the bottom into *valves*, as in Tea or Camellia, and Cotton ; or but partially from the top into *teeth*, as in Chickweeds and Pinks ; or by little openings called pores, such as are found in the ripe fruit of Poppy—is called a *capsule.* And this name is applied to a great variety of fruits, differing much in size and mode of dehiscence, but all agreeing in being syncarpous, and, when quite ripe, dry and dehiscent.

Syncarpous fruits, on the other hand, which are dry and *indehiscent*—that is, which do not open, but liberate the seed by decay, as the fruit of the Oak or of the Buckwheat (*Polygonum*)—you may simply call *nuts.*

In Ranunculus a number of distinct carpels collectively form the fruit, which consequently we have called apocarpous. Each carpel is dry, one-seeded, and indehiscent. Such fruit-carpels are called *achenes.*

FIG. 80. Collective Fruit of Mulberry.

With regard to the Mulberry, the fruit of the Mulberry-tree, you have here the produce not of a single flower, but of a short, dense spike of pistillate flowers, each flower consisting of a perianth of four leaves in two pairs, enclosing the pistil, which is superior, and crowned by a bifid stigma. Now, as the pistil ripens and the seeds mature, the persistent perianth-leaves become very succulent and juicy; and it is to these organs, thus altered in texture, that the Mulberry owes such value as it possesses as an eatable fruit. The Mulberry, therefore, differs from all the so-called berries which we have examined as yet, in

 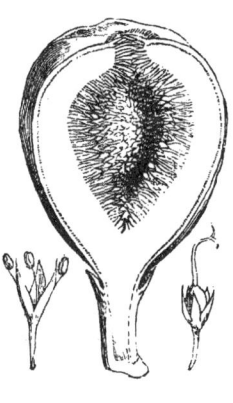

Fig. 81. Jack-fruit (*Artocarpus integrifolia*).

Fig. 82. Fruit of Fig (*Ficus Carica*) in vertical section. Staminate and pistillate flowers are shown separately.

the circumstance that it results, not from a single flower, but from a number of flowers. On this ground it may be distinguished as a *collective* fruit. All fruits which result from more flowers than one are called collective fruits. We have examples in the *cones* of the Pine and Deodar, in the Pineapple, the Jack-fruit, and the fruit of the Fig.

A Fig you can easily get for examination. If cut across,
it appears to be filled with small dry seeds enclosed in a
succulent pericarp. But such is not really the case. Figure
82 shows the staminate and pistillate flowers of the Fig.
In order to observe them you must gather a Fig while
young and green. You will then find that the inside of it
is thickly crowded, not with ovules, but with these minute,
monochlamydeous flowers; the pistillate flowers usually
occupying the lower and greater part of the cavity. It
follows, therefore, that the pulpy portion which forms the
mass of the Fig is a common receptacle, deeply concave
and nearly quite closed at the top, bearing numerous flowers

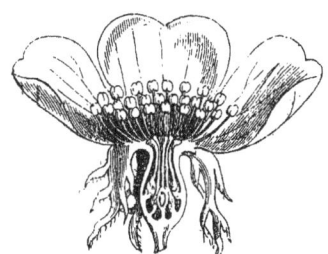

FIG. 83. Vertical section of Flower of Rose, showing the carpels enclosed
in a deeply concave receptacle.

upon its surface. If you have the opportunity, compare
with the Fig the fruit of a Rose. Although rather similar
at first sight, they are essentially different. The fruit of the
Rose results from a single flower, the receptacle of which
becomes more or less succulent and usually red when ripe.
Inside are the separate, dry achenes, which must not be
mistaken for seeds. The fruit of the Rose is analogous to
that of the Strawberry, chiefly differing in the receptacle,
which is concave instead of convex.

Besides the forms of fruit which we have enumerated,

there are a few others so distinct in character as to merit special names and descriptions ; but as these are confined to certain groups of plants, they may be suitably noticed when we come to speak of the general character of the respective groups in Part II.

The prevalent forms of fruit, the usual structure of which should be thoroughly understood, are as follow :—

SIMPLE FRUITS (resulting from a single flower) :

Achene, apocarpous, dry, indehiscent, usually one-seeded (Clematis, Ranunculus, Rose).

Follicle, apocarpous, dry, dehiscing by the ventral suture (Star Anise).

Legume, apocarpous, dry, dehiscing by both sutures (Pea, Indigo).

Nut, properly syncarpous, dry and indehiscent, the peri-carp usually hard and bony.

Drupe, usually apocarpous, succulent, indehiscent and one-seeded, with the inner layer of the pericarp stony (Mango, Almond, Jujube).

Berry, syncarpous, succulent, indehiscent, few or many-seeded (Coffee, Cape Gooseberry (*Physalis*), Grape).

Capsule, syncarpous, dry, dehiscent (Cotton, Lily, Camellia, Rhododendron, Trumpet-flowers).

COLLECTIVE FRUITS (resulting from two or more flowers).

24. The structure of the SEED we have already carefully examined in the Pea, Castor-oil, and Wheat. In examining plants, it will be sufficient at present to note whether the seeds are *solitary, definite,* or *indefinite* in the ovary if it be apocarpous, or in each cell of the ovary if syncarpous. Thus the seeds are *solitary* in Clematis, Cocoa-nut, Mango, Zinnia, Sweet Basil, and Grass-cloth Nettle ; *definite* (few and

constant in number) in the Elephant Creeper and most of
the Convolvulus order ; and *indefinite* (numerous or variable) in Cotton, Gourd, and Pomegranate.   Note also
whether the seeds are *exalbuminous,* that is, containing
embryo only, as in Pea, Mustard, Mango, and Zinnia ; or
albuminous, containing albumen along with the embryo, as
in Coffee, Castor-oil, and Wheat.

CHAPTER VIII.

THE MINUTE STRUCTURE AND VITAL PROCESSES OF PLANTS.

1. IN our second chapter we inquired very briefly into the functions of the nutritive organs. Now that we have had the opportunity of comparing the corresponding organs of many plants, and of forming some tolerable idea of the extent to which the same organ may vary in external

character in different plants, it may be worth while to examine more closely than it was at first expedient into their
mode of working. In order to understand this, you must first
acquire a correct knowledge of the minute composition of
the various organs. Now, their minute composition is,
generally speaking, so simple that you need find no difficulty
in comprehending it; but the parts of which I have to
speak—which build up the leaves, and stem, and root—are
so very minute, that unless you make use of a microscope
that will magnify, say 40 to 80 diameters, you will be unable
satisfactorily to see the parts which compose these organs.
In order to meet this difficulty, in case you cannot get a
sight of the objects themselves, which is always best, refer
to the cuts, which correctly represent all that is necessary.

2. Take first, if you please, a little morsel of the succulent stem of a garden Balsam, or of the stem of any Gourd
or of Water Melon. Boil it for a few minutes, until soft
enough to be torn or dissected out with needles. If you
have none of these plants at hand, a bit as large as a small
pea of any soft herbaceous plant will do. Balsam and
Gourd are particularly well suited, because the parts which I
wish you to examine are not quite so minute in their rapidly
developed stems as they are in plants of less succulence
and firmer texture.

We will suppose that you have taken a very small morsel
of boiled Balsam or Gourd. You observe that it is quite
soft and pulpy, and that a few fibrous strings appear to be
mixed up through it. Take a little of the pulp on the end
of a needle and put it upon a slip of glass, adding a drop
of water. If you have a thin glass cover, put it over the
drop, gently letting one side rest first on the slip as you put
it down, so as to push out the air-bubbles, which are apt to
get entangled, and which look like round balls with black

sides when magnified. When you look at the preparation under the microscope, you are pretty sure to find a number of bodies resembling those represented in the cut. If you do not find them, try another morsel until you succeed. These bodies are called *cells*. They are hollow sacs, each filled with fluid. Now, of cells more or less like these, differing principally in size, in relative length and breadth, and in the thickness of their sides, *every part* of *every plant* is composed. All the organs are built up of these minute cells.

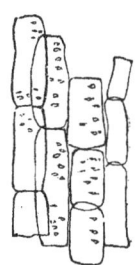

Fig. 84. Cellular tissues; the individual cells oblong, and arranged in vertical rows.

3. But take now a very small bit of one of the fibrous strings of the boiled stem. Place it in a drop of water, and, with a couple of needles, one in each hand, separate it into what seem, to the naked eye, to be its constituent fibres, just as you would separate a morsel of string into its finest threads. When you have got it dissected out, put a cover on as before, and examine it under the microscope. You will probably find here, besides numerous cells of various lengths, some long tubes, with their sides (*walls*) curiously marked with delicate fibres, usually arranged in a spiral direction, twisting round and round inside the tube—the coils sometimes very close, sometimes loose ; or you may find the fibre in the form of separate rings in the inside of the tubes. These tubes are called *vessels*. They originate in this way : A number of cells, such as we saw before, standing one over the other in a row, have the partitions which separate them more or less completely removed, so that the row of cells becomes open all through. We have then a true vessel. Vessels are almost invariably marked either by a spiral, netted, dotted, or ring-like thickening

upon the inside of their walls. In Balsam this thickening usually takes the form of a spiral fibre, but if you boil a bit of Teak or Oak wood, half the size of a pea, in a few drops of nitric acid for a few seconds, it will become white and soft, and after washing it in water two or three times to remove the dangerous acid, you may dissect it in the same way as you did the Balsam. You will find the vessels which it contains more or less like those in Fig. 86. The larger vessels of Teak and Oak wood differ from those of Balsam merely in the thickening on the inside of the vessel

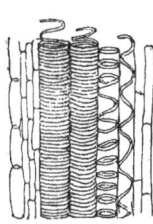
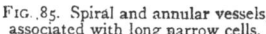
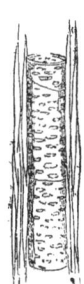

FIG. 85. Spiral and annular vessels, associated with long narrow cells.

FIG. 86. Dotted duct, associated with long, tapering, thick-walled cells.

being more uniformly spread over the wall, omitting only minute spots, which look like holes or pores through the wall of the vessel. Such dotted vessels are very common in wood, and may be easily observed by making very thin slices of the wood lengthwise with a sharp razor.

Plants, then, are built up of cells, or of cells and vessels; the latter originating from cells.

4. Compare with the structure of Balsam that of the following tissues, selected as well suited to show different modifications of cells and vessels, because easily obtained and requiring little preparation.

Pulp of any ripe fruit.   Large, thin-walled cells.

Pith of a young branch of any tree.   The cells are often closely packed, and consequently polygonal.

Piece of the stone of a stone-fruit, or shell of a nut, as Cocoa-nut, ground excessively thin, by rubbing it with the finger upon a hone.   The cells have very thick sides, so

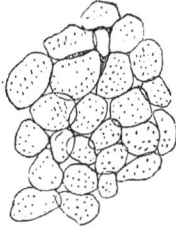

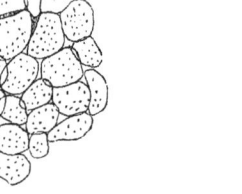

FIG. 87. Thin-walled cellular tissue of pith.

FIG. 88. Cellular tissue, with the walls of the cells much thickened, as in the stone of stone-fruits.

thick that sometimes the cavity is almost obliterated.   The " pores " are lengthened out into long " canals," which radiate from the cavity of each cell.

Yam.   Potato.   Thin-walled cells.

Young shoots of any Fern (boiled to a pulp).   Large vessels, marked with cross bars.

Pine-wood (thin slices, both lengthwise and across the " grain ").   Long, thick-walled, tapering cells, without any vessels.   The sides of the cells are marked with minute disks.

Thin petals, and petals doubled back to show the projecting cells on the folded edge.

Pollen.   The grains are usually oval or roundish, and generally separate.   Compare the pollen of Hibiscus, Cucumber or Gourd, Passion-flower, Lily, any Convolvulus, and

Thunbergia. Examine also the pollen-masses of Orchidaceæ and Asclepiadeæ referred to in Part II.

5. We have spoken of cells as containing fluid. So they
do, as long as they continue to take part in the work of the
plant. But in the trunks of trees the cells forming the
older wood sometimes become so very thick-walled that
they cease to do any work, and, indeed, may be said to
have no room left to do it in.

6. Take some active, sufficiently transparent cells, such
as you find upon the margin of any young leaf bearing short
hairs upon its surface, and, removing a morsel of the leaf
without injuring the hairs upon it, place it in a drop of
water upon a glass slide under the microscope. You observe that each hair is simply a cell of the surface of the
leaf which has grown out into the air. Now, if you add
some fluid that will *kill* the cell, such as a drop of spirits of
wine, you will find, after allowing it time to act, that the
contents of the cell separate from the wall of the cell and
collapse, lying as a loose sac or irregular mass in the middle.
We may, therefore, distinguish cell-contents from cell-wall.
And the distinction is an important one, since all the real
work of the plant is done by the cell-contents; the cell-walls
forming merely the framework of the workshops in which all
the secret and wonderful operations of plants are carried on.

It is this comparative isolation of an infinite number of
vital fragments that constitutes one of the most essential
differences between the vegetable and animal series of the
organized world.

7. In our second chapter we spoke of the elements
carbon, oxygen, hydrogen, and nitrogen, as existing in plants
in a series of peculiar combinations, some of which chemists
are not yet able to imitate in their laboratories. These
combinations we called ternary and quaternary, from their

consisting respectively of three or of four elementary or simple substances. The cell-wall consists of carbon, oxygen, and hydrogen, forming a ternary compound (*cellulose*). The essential part of the cell-contents consists of the same elements combined with nitrogen, forming a quaternary compound. Wherever we have growth going forward, there we have this quaternary compound in activity.

8. The way in which growth in plants takes place is simply this. The contents of the cells of the growing part divide into two, and between the halved contents there forms a thin layer of the ternary cell-wall, which divides each cell into two distinct cells. The new cells then increase in size until they become as large as their parent cell, when they each divide again, and the process is repeated. The process is modified according as the cells are to lengthen or to remain short.

9. In observing the tissue of a Yam or Potato, referred to above as well adapted to show thin-walled, closely packed cells, or of any similar farinaceous tuber or rhizome, you may notice that the cells which are not cut into (and thus emptied) in making a very thin slice are filled with very minute oblong granules. If the slice be too thick, the granules are so numerous that they entirely conceal the delicate cell-walls. These granules, which are stored up in nearly all the cells, are called *starch* granules. To compare with potato-starch, you may take the smallest possible pinch of dry arrowroot and dust it upon your slide, and you will find that the granules of which arrowroot consists, though they differ a little in form, are, in other respects, like those of potato-starch. You may make quite sure of it if you add a small drop of weak tincture of iodine, when they ought at once to become a deep violet; for iodine forms with starch a violet-coloured compound.

Similar granules to those of the Potato, allowing for differences in size and form, you may find in nearly all flowering plants. They are especially abundant in thickened roots, in underground stems, and in seeds. In these organs the starch is stored away as a temporary reserve, to be made use of after a winter's rest, or (in the seed) at the time of germination. The granules then dissolve, and may be said to be eaten by the quaternary cell-contents. Starch is identical in chemical composition with the ternary substance of which the cell-wall is formed. It differs from it in being a temporary deposit instead of a permanent one.

Another form in which temporary reserves are stored up in the cells we find in the globules of oil, abundant in some cells, especially of certain seeds and fruits. Hence we find the principal source of our vegetable oils in the fruits of different plants, as Sesamum (embryo), Olive (pulp of drupe), Cocoa-nut (albumen), &c.

Sugar is another food-deposit of plants, differing from starch in being soluble in the watery cell-sap which fills the full-grown cells. It would appear to be usually an intermediate condition of the plant's ternary food, either preceding the formation or immediately following the solution of starch. It abounds in the cell-sap of the stem of the Sugar Cane and of some Palms.

10. There is another important cell-content of which we have not spoken. If you take any green part of a plant (and it will be best to take a morsel from some succulent leaf, or the thin leaf of a Moss), and examine the cells under a powerful microscope, you will find that the entire cells are not coloured green, neither are the whole of the cell-contents, but that the colouring matter is limited to very minute granules lying in the colourless fluid contents. These are called the *chlorophyll* granules. The development

I

of the green colour of these granules is determined by the action of light, as may be proved by growing plants in total darkness, when they become blanched. The granules are not wholly composed of colouring matter, for the green colour may be easily removed by a little spirits of wine, leaving the granules, which are of quaternary composition, almost unaltered.

11. Besides starch, oil, and chlorophyll, there may frequently be found minute crystals, either in the form of needles, or collected into nodules, lying in the cavity of cells. They are called *raphides*, and are, generally, of subordinate importance.

12. Besides the distinction which I have pointed out between cells and vessels, you may roughly group the different kinds of cells as *long* or *short*. Long cells are usually thick-sided, and often taper at each end, so that when a number of them are grown together, forming a tissue, we find such tissue to be generally firm and tough. Such cells, together with a few vessels, form the principal mass of wood, of petioles, and of the *veins* of leaves. These veins, which have nothing in common with the veins of animals, serve as a sort of framework for the support of the short cells which occupy their interstices. The short cells of leaves are generally thin-walled, and during spring and autumn they are busily engaged in elaborating the food of the plant by the aid of the sun's light and heat. The bundles of long, thick-walled cells, with the vessels which accompany them, forming the veins, we may speak of as the *fibro-vascular* system, and the short cells as the *cellular* system of the leaf. In the petiole the cellular system is much reduced, and the fibro-vascular system is contracted into narrow compass.

13. The arrangement of these systems, as they are termed,

in the stem, differs considerably in the two great Classes of flowering plants.

Excepting in their single cotyledon and the behaviour of the radicle in germination, Monocotyledons are not, at first, materially different from Dicotyledons; but when one or two seasons of growth are over, a marked difference in the mode of arrangement of their fibro-vascular bundles becomes apparent. And this difference essentially consists in the circumstance that in Monocotyledons the fibro-vascular bundles remain permanently isolated, and, once completed in the stem, do not receive any addition in thickness, while in Dicotyledons they become confluent, forming a continuous ring around the pith, and constantly increase in thickness during the successive working seasons of the tree by organically continuous additions to their outer side : so that in Monocotyledons the bundles are *closed* or *definite;* in Dicotyledons, *continuous* or *indefinite.*

But the nature of this difference you will appreciate better when you understand the composition or arrangement of the tissues forming these fibro-vascular bundles. Each bundle contains at first a layer of cells of extreme delicacy, which cells are capable of undergoing division and enlargement ; and it is by means of this layer only that the bundle can increase in thickness. This layer of active cells is enclosed between two distinct systems; one system, on the side towards the centre of the stem,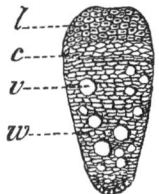

Fig. 89. Diagram representing the arrangement of the tissues in a fibro-vascular bundle; *l* the liber, *c* cambium-layer, *w* wood, *v* wide vessel of the wood.

consisting of long, thick-walled cells and vessels, forming the proper *wood* of the fibro-vascular bundle, and another (smaller) system, on the side of the bundle towards the circumference of the stem, characterised by thick-walled, tapering cells, forming the *liber* system.

The figure (89) represents a fibro-vascular bundle cut across, showing at *c* the layer of delicate cells, called the *cambium-layer*, the cells of which divide and give off new cells on each side—on the inner side wood-cells (*w*) an vessels (*v*), on the outer side fibrous liber-cells (*l*). In Dicotyledons these bundles are arranged in the stem in such a manner, at a very early stage of its growth, that the cambium-cells of the bundles, which are side by side, coalesce, and thus form one continuous cylinder of multiplying and enlarging cells. The consequence is, that in Dicotyledons all the *wood* is on the inside of this cambium-

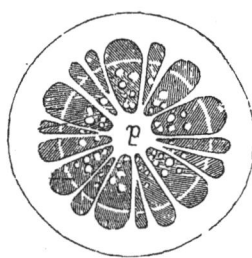

cylinder, and new wood is deposited on the outside of wood previously formed ; all the *liber*, on the other hand, is on the outside of the cambium, and immediately within the bark—of which, indeed, the liber is regarded as forming an inner layer. Structure such as here described is precisely what we find in the stem of a Mango, or any other dicotyledonous tree, which is said to be *exogenous*, from the circumstance that the wood increases by additions to its outside.

FIG. 90. Diagram showing the arrangement of the fibro-vascular bundles (each of them wedge-shaped in cross-section) in the stem of a young Dicotyledon. The pale circle passing through each bundle near its thicker end indicates the future cambium-cylinder; *p* the pith.

In Monocotyledons, on the other hand, the cambium-cells of the different fibro-vascular bundles never coalesce so as to form a cambium-cylinder: consequently they do not form continuous rings of wood. The cambium-cells, therefore, soon cease their dividing and enlarging work, and the fibro-vascular bundle is finished. We find, if we cut the solid stem of a Monocotyledon across, that the fibro-vascular bundles are irregularly scattered all through the cellular system of the stem. They are especially crowded towards the circumference, which consequently becomes much harder than the centre in Woody Monocotyledons. From the mode of development of the fibro-vascular bundles, and the direction which they take in the stem, the trunk of Woody Monocotyledons does not usually increase in diameter

FIG. 91. Diagram showing the arrangement of the fibro-vascular bundles in a cross-section of the stem of a Monocotyledon. *v b* fibro-vascular bundles ; *c t* cellular tissue.

beyond a certain point, as we find in Palm-trees, which frequently have tall cylindrical stems as thick at the top as at the base. These peculiarities led the older botanists to call such stems *endogenous,* from a notion that the younger bundles were those in the centre of the stem, and that they pushed and compressed the older bundles towards the outside.

14. But the structure of Woody Dicotyledons requires further examination.

Take a cross-section of the stem of an Oak, or of any branching tree from the cooler mountain ranges, already several years old. You find in the centre the remains of the original cellular system of the stem reduced to a very narrow cord, and distinguished as the *pith.* When young the pith served to contain and to convey nourishing fluids to the growing point; now it is dry and useless. Surrounding the pith is the wood, forming the great mass of the stem. A number of concentric rings are distinguishable in the wood, there being as many rings as years that the stem has existed, one ring to a year; so that by counting the rings you may ascertain the age of the stem. The appearance of rings, or *annual zones*, in the wood, arises simply from the wood formed in summer and autumn being denser, closer grained, and with fewer vessels than that formed in spring.

With a little care you may notice that there are, as it were, narrow rays proceeding from the pith to the bark. These are actual plates of cellular tissue left between the fibro-vascular bundles, which look like narrow rays when cut through transversely. They are called the *medullary rays*. They are usually very narrow, much narrower in most trees than in the Oak, the wood of the common European species of which, when cut lengthwise in the direction of these rays, is marked by silvery patches of the cells of the rays, forming what is called the silver-grain, which painters imitate in painting wainscot.

It will be difficult to find the cambium-ring without using a lens, but it is immediately within the *bark*, which it connects with, while at the same time it separates it from, the wood.

The outer layers of bark are usually composed of short cells of corky texture, which serve to prevent the cambium-layer from drying up, by checking evaporation from the surface.

The inner or liber-layer of the bark of many plants and trees is made use of for cordage and in cloth-making. Hemp, sunn-hemp, jute, and flax, are all derived from this layer, which, in the plants affording these products, is very tough.

15. The questions now present—Through which of these cells, or systems of cells, in the stem, is the watery sap absorbed by the roots conveyed to the leaves? And when

the sap has been exposed to the sun's influence in the
leaves, how does it find its way through the plant? In
other words, How do the sap and nutrient fluids circulate in
the plant? But these questions, reasonable though they
seem, it is impossible to answer satisfactorily in the present
state of our knowledge.

It is not our business just now to concern ourselves with
contested points; so we must be content with a very general
and partial explanation. In the first place, we must recall
the fact that the entire plant is built up of closed cells and
vessels ; consequently, solid substances, even in the minutest
state of subdivision and suspension in water, cannot be
admitted by healthy uninjured plants. Mistakes as to this
point have arisen from using injured or wounded plants for
experiment. Therefore only fluids, substances dissolved in
fluids, and gases, can be absorbed by the plant ; viz. fluids
with solids or gases in solution by the roots, gases and
vapour by the leaves.

The circulation or transference of these fluids and gases
from cell to cell can only be by *diffusion*, a physical process,
probably controlled, in some way not yet understood, by its
taking place in a living apparatus. This process of diffusion
depends upon two conditions. First, we must have two
fluids separated by a membrane of some kind which they
can permeate. Second, these fluids must be of different
chemical composition, or of different density. When these
conditions exist, a current is set up through the membrane,
which results in one of the fluids (the denser) increasing in
bulk at the expense of the other. This increase is due to
Diffusion. The affinity of the membrane itself for one of the
fluids in preference to the other modifies the result. Now
these conditions obtain throughout all plants, excepting, of
course, the old dead and dry portions of trunks, &c. They

are built up of closed cells, containing fluids of various density, and the walls of the cells are permeable. The consequence is, that there is a constant transmission of fluids going forward throughout their tissues. The direction of the current is mainly determined by the constant evaporation from the leaves, which necessarily tends to render their cell-contents denser, so that the water taken up by the surface-cells and hairs of the root-fibrils is impelled upwards cell by cell, to restore the equilibrium, until it reaches the leaves and other tissues exposed to evaporation. The course which the *ascending sap,* as it is termed, takes has been usually supposed to be through the cells forming the younger layers of wood, the vessels assisting when the current is rapid. This, however, is not absolutely determined. The *elaborated sap* (that is, the sap after having undergone certain chemical changes, especially in the leaves, determined by the influence of the sun) is generally admitted to descend chiefly through the inner layers of bark. Indeed, a rude experiment may be regarded as strongly confirming this view. If you remove a ring of bark from the stem of a tree, or bind it very tightly round with a strong hoop, no wood will be formed below the ring or hoop. On the other hand, a considerable thickening will take place immediately above it.

From the absence of a system of vessels analogous to that of animals, and of a pumping-engine like the heart, the course taken by fluids in plants is comparatively very vague and ill-defined at best. I have here merely indicated its general course in the stems of Dicotyledons.

16. I have spoken of leaves as capable of absorbing gases, especially carbonic-acid gas, and probably also vapour, from the atmosphere.

If a leaf be examined carefully, it will be found covered

with a thin skin or epidermis, which very often (in fleshy leaves) may be torn off in filmy shreds.   And a similar epidermis covers nearly all the green and coloured organs which are exposed to the air.   If a piece of this epidermis, torn from a leaf with the thumb and a sharp penknife, be placed in a drop of water upon a glass slide, its structure may be easily made out under the microscope. Suppose a shred torn from the leaf of a Lily.   It will be found to consist of an excessively thin layer of flattened cells, closely fitting at their angles.   Scattered at intervals over the epidermis are pairs of very small cells side by side, with their ends in contact, as shown in the cut.   Each pair of cells forms a *stomate*.   When the cells of the stomates are rendered turgid by the absorption of fluid, they separate

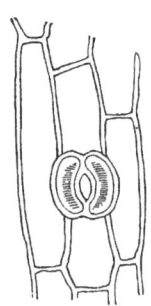

Fig. 92. Fragment of Epidermis with a single stomate.

more or less from each other, leaving a minute opening in the middle between them.   When they are flaccid, the guard-cells remain closely applied, and the orifice is closed. Under ordinary conditions of the air as to moisture they are open ; when it is either very dry or very moist, they are generally closed.

The stomates, therefore, serve to facilitate the absorption of gases, and probably of vapour, from the air.   They do not, however, open into cells, but into spaces between the cells of the leaf, called *intercellular spaces*.   These intercellular spaces are widest between the cells forming the lower layers of the leaf, and we find that stomates are generally much more abundant in the epidermis of the lower than of the upper surface of leaves.   There are no stomates on roots, nor, usually, on surfaces under water.

PART II.

FIRST BOOK OF INDIAN BOTANY.

CHAPTER I.

CLASSIFICATION OF PLANTS.

1. Extension of the plan of examining Type-specimens to subordinate groups.
2. The specific and generic names of plants.   Individuals.   Species. Genera.
3. The Binomial method of naming plants.   Diagnostic characters.
4. Arrangement of genera under superior groups.   Subordination of characters.
5. Varieties.
6. Explanation of the Type lessons.   Necessity for a constant reference to specimens.

1. IN preceding chapters I have endeavoured to illustrate the prevalent structure of Dicotyledons and Monocotyledons, by referring to a very limited number of common plants, which we made use of to illustrate different kinds of modification in the various organs, and especially in the parts of the flower.

Thus we made use of the Poppy, and others, as examples of the Dichlamydeous Sub-class, the Grasscloth Nettle of the Monochlamydeous Sub-class, and the Willow of the Achlamydeous Sub-class of Dicotyledons.   And further, we employed

Poppy, Mustard, Rose, and Melastoma as examples of the Polypetalous Division, and Zinnia, Rose, Periwinkle, and Sweet Basil of the Gamopetalous Division, of the Dichlamydeous Sub-class ; and so on for each of the principal divisions of Dicotyledons and Monocotyledons.

These illustrative examples we may regard as representative Types.   Each type embodies the characteristics of a large group, the members of which group, though they differ from each other in minor details, such as regularity or irregularity of the corolla, and sometimes in the number of stamens and of carpels, generally agree in characters which, from experience, we infer to be important, from their prevalence through a large number of plants.   These important characters are principally based upon adhesion, cohesion, and suppression of the parts of the flower.

By extending this method, by selecting and carefully studying types representing the principal subordinate groups, called NATURAL ORDERS, of Indian Flowering Plants, we shall lay the sure basis of a thoroughly scientific acquaintance with them.   The types which we presently proceed to select from each important Natural Order are not always the best suited to serve as representatives of such Order, because we shall be obliged to make use of plants of which specimens may be easily obtained, and these do not always happen to be best adapted for the purpose.   Besides, in some Natural Orders the amount of variation in minor characters is so considerable that we shall find it needful to employ Sub-types, the relation of which to their type we shall endeavour to make clear whenever we find it needful to employ them.

You must not be content with the examination of those plants only which are employed as types.   You must try to refer to its type *every flowering plant you meet with*, and, in a short time, you cannot fail to recognise easily the Natural

Orders to which most Indian plants belong. In the follow-lowing pages you will observe that in nearly every case each plant is designated both by an English or native and by a scientific name. This is done partly that you may be familiarized with a plan of naming plants based upon definite principles, and partly that the memory may be stored (though we would not have it burdened) with at least the *generic* scientific names of the more important common native and cultivated plants, which names are in use amongst botanists of all countries.

2. The scientific name of every plant consists of two words, a substantive and an adjective. The substantive is the name of the *genus*, as Brown or Jones may be the name of a family. The adjective indicates the *species*, as John, Thomas, or William indicates the individual member of a family.

But *species* is a collective term, and the same *specific name* is applied to all the *individuals* which belong to the same species. All individual plants which resemble each other so nearly that it is consistent with experience to suppose that they may all have sprung from one parent stock, are regarded as belonging to the same species. In other words, the differences between the individuals of the same species are generally not greater than we are accustomed to observe between the individual plants in a field of Poppies, or of Rice, or in a bed of any garden annual sown with seed which we know to have been gathered originally from a single plant. All plants, therefore, which resemble each other thus nearly are referred to the same species, and the same specific adjective name is employed to designate them.

Then again, species which resemble each other in all important particulars of structure (though it is impossible to define the exact particulars, for to a great extent they are arbitrary and of convenience,) may be referred to the same *genus*, and the same generic substantive name is employed to designate them. Thus, we refer all the species of Balsam to the genus *Impatiens*, and of Fig to the genus *Ficus*. In this way we have *genera* (plural of genus) including often many species, sometimes several hundreds; we have others, again, which include few or but single species. In the latter case we have species which are necessarily comparatively isolated in the characters of their flowers; more so, at least, than are the species of larger genera.

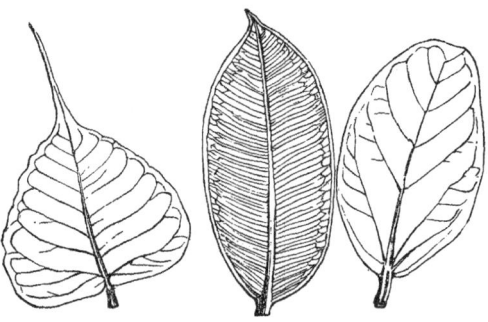

FIG. 93. Leaf of Peepul   FIG. 94. Of India-rubber   FIG. 95. Of Banyan
(*Ficus religiosa*).   Fig (*F. elastica*).   (*F. benghalensis*).

3. Recollect, then, that in the scientific name of a plant we always state both the name of the genus and that of the species to which it belongs. The generic name precedes. Thus *Ficus benghalensis*, the Banyan, *Ficus religiosa*, the Peepul, and *Ficus elastica*, the India-rubber tree, are three distinct species of Fig familiar to every one in India; and although they resemble each other very closely in the structure of their flowers and seeds, yet inspection of a single detached leaf will enable any one to distinguish them. The brief characters which suffice to distinguish these species from

each other are said to be *diagnostic.* The diagnostic characters, derived from the leaves only, of these three species
of *Ficus* are:

F. benghalensis—Leaves elliptical-ovate, obtuse; nerves
prominent beneath, distant, with intermediate reticulation,
the lower palmate.

F. religiosa—Leaves broadly ovate, narrowed into a long
slender acumen, base narrowly cordate; nerves pinnate
from the base, scarcely prominent beneath, with intermediate
reticulation.

F. elastica—Leaves oblong or elliptical, abruptly apiculate,
base entire; lateral nerves obscure, pinnate from the base,
very numerous and closely parallel, without intermediate
reticulation in the lower part of their course.

It will be observed that the diagnoses of species rest upon
comparatively slight modifications of structure. The diagnoses of genera rest upon characters of higher importance
(characters more constant in the group than those used to
distinguish species), and so on for the groups superior to
the genus, the characters of each of which embrace, as we
have already shown, those of all their subordinates. The
method of denoting every plant and animal by two names, a
generic and specific, on a uniform plan, was invented by
Linnæus, and this Binomial method is now universally
adopted by naturalists.

The method of grouping genera into higher groups, according to their resemblance in characters of successive
degrees of constancy, though indicated by the same eminent
man, has been the work of many collaborators, amongst
whom the names of Jussieu, Brown, and De Candolle are
pre-eminent.

4. Precisely as we group species under Genera, so we
group genera under Natural Orders. The Natural Orders

again (to which substantive names are applied for convenience) under Divisions, the divisions under Sub-classes and Classes, as we have already pointed out. Thus the characters of a Class are common not only to its Sub-classes and Divisions, but to the Natural Orders, Genera, and Species included in that Class. It follows, therefore, that the characters of a Class must be more constant and more general than those of a Sub-class or Division, those of a Division than those of a Natural Order, of a Natural Order than those of a Genus, and of a Genus than of the Species which it includes.

5. Botanists distinguish as *varieties* groups of individuals of a species which are marked in common by some trivial character, subordinate in importance to the characters which are used to separate species. Thus we may have orange and purple varieties of the same species of Zinnia, awnless and awned varieties of the same species of Rice, &c.; the colour of the flower of the Zinnia and the presence or absence of an awn in Rice being characters too liable to variation to serve to separate species.

6. The following pages are chiefly devoted to an examination of representative Types of most of the Natural Orders of flowering plants native in India.

I must here emphatically impress upon the beginner, that it is useless attempting to study this portion of the book without a constant reference to living specimens, without which any information he may acquire from it will be comparatively unavailable when tested in the field. Numerous references are given to plants which show peculiar departures from the several Types. Specimens of these ought to be procured whenever it is possible, and dried for further use in the way described in the last chapter of this book. When a preparation can be preserved without pressing it between papers, as for example many dried fruits, seeds, galls, spines, &c., it would be well to have them thoroughly dried and mounted upon pieces of card, labelled with the name of the plant, the Natural Order to which it belongs, the particular in which it departs from the Type, &c. Preparations of plants used for economic purposes, whether domestic, medicinal, in the arts, or otherwise, are always interesting, and are very useful for purposes of illustration. A few of these, which may be easily obtained, I have indicated; but there are hundreds not mentioned and equally accessible.

CHAPTER II.

DICOTYLEDONS.

DICHLAMYDEÆ. — Perianth double (calyx and corolla usually both present).

POLYPETALÆ.—Corolla polypetalous (excepting in many Ternströ-miaceæ and Olacaceæ, and in isolated genera of other Natural Orders).

THALAMIFLORÆ.—Stamens usually hypogynous.

* *Pistil apocarpous* (excepting in some Dilleniaceæ, Nymphæaceæ, and in the genus Anona).

RANUNCULACEÆ (p. 146). — Herbs with radical and cauline (rarely floating) alternate leaves, or climbing shrubs with opposite leaves (*Clematis*). Petals five, fewer or none ; if none, the sepals are petal-like. Sepals and petals deciduous. Stamens usually indefinite. Seed albuminous, without an aril.

DILLENIACEÆ (p. 148).—Trees or shrubs with alternate leaves, or herbs with radical leaves (*Acrotrema*). Sepals persistent. Petals five or four. Stamens indefinite. Seeds with an aril.

MAGNOLIACEÆ (p. 149).—Trees or shrubs with convolute stipules (except *Illicium*, an exstipulate evergreen shrub of

Eastern Bengal). Sepals and petals trimerous (except *Illicium*), imbricated. Stamens indefinite. Albumen uniform.

ANONACEÆ (p. 151).—Trees or shrubs with alternate entire exstipulate leaves. Sepals and petals trimerous. Stamens indefinite. Albumen ruminated.

MENISPERMACEÆ (p. 153). Climbing or twining shrubs (except in *Cocculus laurifolius* of the Himalaya, a small tree or erect shrub), and very small diœcious, usually trimerous flowers. Stamens definite, often monadelphous. Carpels usually three, varying with some multiple of three, or with one. Embryo usually curved.

NYMPHÆACEÆ (p. 155).—Aquatic herbs, usually with large flowers. Petals and stamens indefinite. Stigmas distinct.

** *Pistil syncarpous* (except in Connaraceæ).

PAPAVERACEÆ (p. 157).—Herbs with white, coloured, or (in Tribe *Fumarieæ*) watery juice. Stamens indefinite, free, or (in *Fumarieæ*) definite, diadelphous. Ovules parietal.

CRUCIFERÆ (p. 159).—Herbs with alternate exstipulate leaves. Sepals four. Petals four. Stamens usually tetradynamous. Seeds without albumen.

CAPPARIDACEÆ (p. 161).—Herbs or shrubs with alternate, often divided leaves. Sepals four. Petals four. Stamens indefinite, or eight, six (not tetradynamous), or four. Ovary often stipitate. Seeds usually exalbuminous.

VIOLARIEÆ (p. 163). — Trees, shrubs, or herbs with alternate stipulate leaves. Sepals five. Petals five. Stamens five ; the connective of the anthers usually dilated, and produced beyond the anther-cells. Ovary one-celled ; placentas three.

BIXACEÆ (p. 164).—Trees or shrubs with alternate simple leaves. Flowers unisexual or hermaphrodite, often apetalous. Stamens indefinite. Seeds albuminous.

POLYGALACEÆ (p. 165).—Trees, shrubs, or herbs with alternate simple exstipulate leaves. Sepals unequal. Stamens usually eight, united below, with the petals adherent to the staminal sheath.

CARYOPHYLLACEÆ (p. 166).—Herbs with opposite simple leaves. Ovary one-celled, with a free central placenta.

HYPERICINEÆ (p. 167).—Herbs, shrubs, or rarely trees, with opposite undivided leaves. Flowers hermaphrodite. Calyx imbricate. Stamens indefinite, frequently cohering below in three or more phalanges.

GUTTIFERÆ (p. 168).—Trees or shrubs with a resinous, often coloured, juice, and opposite undivided glabrous leaves. Flowers unisexual or polygamous. Calyx imbricate.

TERNSTRÖMIACEÆ (p. 169).—Trees or shrubs, usually with alternate simple leaves. Flowers hermaphrodite or unisexual. Calyx imbricate. Stamens rarely definite.

DIPTEROCARPEÆ (p. 170).—Trees or climbing shrubs (*Ancistrocladus*), usually resinous, with alternate penni-veined simple leaves. Two or more of the segments of the calyx-limb usually enlarged in fruit.

MALVACEÆ (p. 173).—Herbs, shrubs, or trees with alternate simple, frequently palmi-nerved leaves. Calyx-lobes valvate. Stamens monadelphous, with one-celled anthers.

STERCULIACEÆ (p. 176).—Trees, shrubs, or herbs with alternate simple or digitate leaves. Calyx-lobes valvate. Stamens monadelphous, indefinite or definite, or free and definite, with or without alternating staminodes; anthers two-celled.

TILIACEÆ (p. 177).—Trees, shrubs, or herbs usually with alternate simple leaves. Calyx-lobes valvate. Stamens indefinite, usually free; anthers two-celled.

LINACEÆ (p. 178).—Shrubs, herbs, or rarely trees, with alternate undivided simple leaves. Sepals imbricate. Stamens definite, more or less coherent below. Ovary undivided; ovules one or two in each cell.

MALPIGHIACEÆ (p. 179).—Climbing shrubs with opposite entire leaves. Calyx often with sessile glands. Petals usually clawed. Stamens ten. Ovary three-lobed. Carpels winged.

ZYGOPHYLLACEÆ (p. 180).—Herbs or low shrubs with opposite stipulate compound leaves. Peduncles axillary, one-flowered. Stamens eight or ten, free, often with a minute scale at the base of the filament.

GERANIACEÆ (p. 181).—Herbs with opposite or alternate simple or compound leaves. Flowers regular or irregular (the posterior sepal spurred in Tribe *Balsamineæ*), usually two or more on axillary peduncles. Stamens definite. Ovary lobed.

RUTACEÆ (p. 182).—Usually trees or shrubs with alternate or opposite compound (pinnate, or tri- or uni-foliolate) leaves, dotted with translucent glands. Stamens as many or twice as many as petals (in *Citrus* and *Aegle* indefinite).

OCHNACEÆ (p. 184).—Shrubs or trees with alternate, shining, coriaceous, simple, eglandular leaves. Anthers linear, often elongate. Ovary lobed deeply. Fruit of three to five, or more, distinct drupes.

BURSERACEÆ (p. 185).—Trees with resinous juice, alternate compound leaves, and small, panicled or racemose flowers. Stamens free, as many or twice as many as petals. Ovary entire, with two or more cells.

MELIACEÆ (p. 186).—Trees or shrubs with alternate compound leaves, and panicled flowers. Stamens definite, monadelphous (except in *Cedrela* and *Chloroxylon*). Ovary entire.

OLACACEÆ (p. 188).—Shrubs or trees usually with alternate entire leaves, and axillary fascicles spikes or racemes of small flowers. Petals free or connate, usually valvate. Ovary one—or imperfectly three—or more-celled.

AMPELIDEÆ (p. 188).—Usually shrubs with jointed stems, climbing by tendrils, or in *Leea* erect. Leaves alternate, simple, or tri- to quinque-foliolate (or compound-pinnate in *Leea*). Flowers minute, cymose, greenish. Petals valvate, caducous. Stamens as many as petals, and opposite to them.

SAPINDACEÆ (p. 190).—Usually trees (in *Cardiospermum*, a scandent herb,) with alternate pinnate leaves, and small, often panicled, polygamous flowers. Stamens free, frequently anisomerous. Ovary three-, four-, or two-celled.

ANACARDIACEÆ (p. 192).—Trees, often resinous, with alternate or opposite, simple or compound leaves. Flowers small. Ovary one-celled (in *Spondias* and allies, two to five-celled) ; ovules solitary.

CONNARACEÆ (p. 193).—Trees or shrubs with alternate compound leaves. Flowers small, regular, in racemes or panicles. Stamens definite. Pistil apocarpous.

CALYCIFLORÆ.—Stamens usually perigynous or
epigynous.

LEGUMINOSÆ (p. 194).—Trees or shrubs usually with alternate compound (pinnate, tri- or uni-foliolate) leaves. Flowers irregular (except in Tribe *Mimoseæ*). Carpel solitary.

ROSACEÆ (p. 200).—Trees, shrubs, or herbs with alternate, entire or divided leaves. Flowers regular. Ovary free or adherent to the calyx-tube (when, if there be two or more carpels, it becomes apparently syncarpous).

COMBRETACEÆ (p. 203).—Trees or shrubs with opposite or alternate simple leaves. Flowers with or without petals. Ovary wholly inferior, one-celled, with pendulous ovules.

CELASTRACEÆ (p. 204).—Shrubs or trees with opposite or alternate simple leaves, and minute flowers. Ovary more or less immersed in a disk. Stamens alternate with the petals and equal in number, or only three.

RHAMNACEÆ (p. 205).—Trees or shrubs with alternate simple leaves, and minute flowers. Stamens opposite to the petals and equal in number.

MELASTOMACEÆ (p. 206).—Herbs or shrubs with opposite entire, usually three-nerved leaves. Petals twisted in bud. Stamens ten or fewer, perigynous; anthers frequently appendaged. Ovary free or adnate.

MYRTACEÆ (p. 208).—Trees or shrubs usually with opposite entire leaves marked with translucent glandular dots. Stamens indefinite, perigynous. Ovary adherent, with axile placentation.

RHIZOPHORACEÆ (p. 210).—Trees or shrubs with opposite entire coriaceous leaves. Calyx-lobes valvate. Petals often fringed. Stamens usually some multiple of the petals. Ovary more or less adherent.

ONAGRACEÆ (p. 211).—Herbs with alternate or opposite simple (sometimes if submerged, divided) leaves. Calyx-teeth usually four, valvate. Ovary inferior, two- or four-celled, with indefinite (rarely definite) ovules.

LYTHRACEÆ (p. 212).— Trees, shrubs, or herbs with opposite or alternate simple leaves. Ovary free.

CUCURBITACEÆ (p. 213).—Climbing or prostrate herbs with alternate, usually palmately-nerved leaves, and lateral tendrils. Flowers unisexual. Stamens usually three (one with half an anther). Ovary inferior. Seeds exalbuminous.

BEGONIACEÆ (p. 216).—Succulent herbs with oblique, radical, or alternate cauline leaves. Flowers unisexual. Stamens indefinite. Ovary inferior, three-celled; ovules indefinite.

CRASSULACEÆ (p. 217).—Herbs or shrubs, usually with fleshy leaves. Flowers regular, hermaphrodite. Stamens definite. Pistil nearly apocarpous; ovary superior.

SAXIFRAGEÆ (p. 218). — Herbs, shrubs, or trees with alternate or opposite simple leaves (in Indian species). Stamens usually ten or fewer, perigynous. Ovary more or less adherent to the calyx-tube. Seeds usually indefinite, albuminous.

UMBELLIFERÆ (p. 219).—Herbs with hollow stems and alternate sheathing, divided or dissected leaves (orbicular and peltate in *Hydrocotyle*), and small umbellate flowers. Petals five. Stamens five; epigynous. Carpels two; when ripe, dry, indehiscent, and usually separating.

ARALIACEÆ (p. 222). — Trees, shrubs, or herbs with alternate simple or compound leaves. Petals and stamens epigynous. Fruit usually succulent, indehiscent, and not separating into its constituent (two or more) carpels.

LORANTHACEÆ (p. 223).—Parasitical shrubs with opposite or alternate coriaceous simple leaves. Stamens opposite to the apparent petals, and adnate to them below. Ovary inferior. Fruit one-seeded. Seed albuminous.

GAMOPETALÆ. — Corolla with the petals united. (Petals exceptionally free in some species of Ericaceæ, Campanulaceæ, Styracaceæ, Oleaceæ, Primulaceæ, and Plumbaginaceæ).

* *Ovary inferior* (except in a few Campanulaceæ and Styracaceæ).

RUBIACEÆ (p. 224).—Trees, shrubs, or herbs with opposite simple leaves and interpetiolar stipules, or in some herbs with verticillate leaves. Stamens as many as corolla-lobes, epipetalous.

COMPOSITÆ (p. 227). — Trees, shrubs, or herbs with alternate or opposite leaves, and capitate inflorescence. Stamens as many as corolla-lobes; anthers syngenesious. Ovary one-celled, with one erect ovule.

CAMPANULACEÆ (p. 231).—Herbs or shrubs with alternate or opposite exstipulate leaves and milky juice. Stamens as many as corolla-lobes; epigynous. Ovary two- to three- or more-celled. Seeds indefinite.

STYRACACEÆ (p. 233). —Trees or shrubs with alternate simple leaves. Stamens often indefinite, inserted on the base of the corolla-tube. Seeds usually solitary.

ERICACEÆ, Tribe *Vacciniea* (p. 232).—Shrubs or trees with alternate simple leaves. Stamens epigynous, usually twice as many as corolla-lobes. Anthers opening by pores, often appendaged. Seeds indefinite.

** *Ovary superior.*

ERICACEÆ, Tribe *Ericea*, (p. 233).—As in Tribe *Vacciniea*.

EBENACEÆ (p. 234).—Trees or shrubs with alternate entire leaves, and regular, usually polygamous, flowers. Stamens inserted on the base of the corolla or hypogynous, usually a multiple of the corolla-lobes. Ovules one or two in each cell of the ovary.

SAPOTACEÆ (p. 235).—Trees or shrubs with alternate entire leaves. Flowers regular. Stamens opposite to the corolla-lobes, as many or twice as many; often with numerous scale-like staminodes. Ovules solitary or in pairs, in each cell of the ovary.

OLEACEÆ (p. 236).—Trees or shrubs with opposite leaves. Flowers regular. Stamens two. Ovary two-celled; ovules one or two in each cell.

APOCYNACEÆ (p. 238).—Trees, shrubs, often climbing or twining, or herbs with opposite or rarely alternate, exstipulate simple entire leaves. Flowers regular. Stamens five, alternate with corolla-lobes. Carpels two, usually distinct in the ovary, united in the style and stigma.

ASCLEPIADEÆ (p. 239).—Shrubs or herbs, often climbing, with opposite entire exstipulate leaves. Flowers regular. Stamens five; anthers coherent around the stigma; pollen usually consolidated in masses attached to minute stigmatic glands. Carpels as in *Apocynaceæ*.

LOGANIACEÆ (p. 242).—Trees, shrubs, or herbs, sometimes twining, with opposite entire leaves, often with interpetiolar stipules. Flowers regular. Stamens as many as, and alternate with, the corolla-lobes. Ovary usually two-celled.

GENTIANACEÆ (p. 243). — Herbs with opposite entire leaves, rarely twining, or an aquatic with floating leaves; taste bitter. Flowers regular. Stamens as many as, and alternate with, the corolla-lobes. Ovary usually one-celled, with parietal placentas and indefinite ovules.

BIGNONIACEÆ (p. 244).—Usually climbing shrubs or trees with opposite, rarely simple, exstipulate leaves. Flowers irregular. Stamens fewer than corolla-lobes. Ovary two-celled, with indefinite ovules. Fruit a capsule, with winged seeds.

PEDALIACEÆ (p. 245). — Herbs with opposite leaves. Flowers irregular. Stamens fewer than corolla-lobes. Pistil usually dicarpellary with the ovary four-celled; ovules indefinite, rarely solitary.

CONVOLVULACEÆ (p. 247).—Herbs, shrubs, or rarely trees, usually twining or prostrate (in *Cuscuta* a leafless parasite), with alternate leaves, and usually showy regular flowers, with a plaited funnel-shaped corolla. Stamens as many as corolla-lobes, and alternate with them. Ovary two- or four-celled, with one or two ovules in each cell. Cotyledons usually folded.

BORAGINACEÆ (p. 248).—Herbs, shrubs, or rarely trees, with alternate simple, often roughly hairy, exstipulate leaves, and cymose, usually unilateral, inflorescence. Flowers regular. Stamens as many as corolla-lobes, and alternate with them. Ovary four- (or two-) lobed, with one ovule in each lobe. Style one, sometimes forked above.

SOLANACEÆ (p. 249).—Herbs, shrubs, or sometimes trees, with alternate exstipulate leaves, and often extra-axillary inflorescence. Flowers regular. Stamens as many as, and alternate with, corolla-lobes (anthers opening in terminal pores in *Solanum*). Ovary two-celled, with indefinite ovules. Seeds albuminous ; embryo rarely straight.

SCROPHULARIACEÆ (p. 252).—Usually herbs with alternate or opposite exstipulate leaves. Flowers irregular. Stamens fewer than corolla-lobes. Ovary two-celled, with indefinite ovules. Seeds albuminous ; embryo usually straight.

LENTIBULARIEÆ (p. 254).—Herbs growing in water or in damp places; the aquatic species usually with dissected leaves bearing minute air-vesicles. Flowers two-lipped. Stamens two. Ovary one-celled, with a free central placenta and indefinite ovules.

ACANTHACEÆ (p. 255).—Herbs or shrubs (in *Thunbergia* usually twining) with opposite simple exstipulate leaves. Flowers irregular, often with conspicuous bracts. Stamens fewer than corolla-lobes. Ovary two-celled, with two or more ovules in each cell. Fruit a two-valved capsule, with the seeds supported on horny hooks or cushions of the placenta.

LABIATÆ (p. 257).—Herbs or shrubs, usually aromatic, with opposite leaves and irregular flowers. Stamens fewer than corolla-lobes. Ovary four-lobed, four-celled, with one ovule in each cell. Style one, gynobasic.

VERBENACEÆ (p. 259). — Trees, shrubs, or herbs with opposite leaves and irregular flowers. Flowers fewer than corolla-lobes. Ovary four-celled, entire with one ovule in, each cell. Style one, terminal.

MYRSINACEÆ (p. 260).—Trees or shrubs, with alternate often glandular-dotted entire leaves, and regular flowers. Stamens as many as corolla-lobes, and opposite to them. Ovary one-celled, with a free central placenta. Seeds one or more.

PRIMULACEÆ (p. 261).—Herbs, with radical or alternate leaves. Flowers regular. Stamens as many as corolla-lobes, and opposite to them. Ovary one-celled, with a free central placenta. Style one. Seeds indefinite.

PLUMBAGINACEÆ (p. 263).—Herbs or shrubs with radical or alternate leaves. Flowers regular. Stamens as many as corolla-lobes (often nearly free to the base), and opposite to them. Ovary one-celled, with a solitary ovule suspended from a slender funicle. Styles five.

PLANTAGINACEÆ (p. 264).—Herbs with radical or rarely cauline leaves, and small greenish or scarious, regular, often spicate flowers. Sepals four. Stamens four, exserted, and alternate with corolla-lobes. Style simple.

INCOMPLETÆ.—Perianth simple or none (double in some Euphorbiaceæ).

* *Angiospermous* (Ovules fertilised through the medium of a stigma).

NYCTAGINACEÆ (p. 265).—Herbs, shrubs, or trees with alternate or opposite (unequal) leaves. Flowers regular, hermaphrodite, usually coloured ; base of the perianth persistent, closely investing the superior one-celled, one-seeded nut. Seed albuminous, with a curved embryo.

CHENOPODIACEÆ (p. 266).—Usually herbs with alternate or opposite exstipulate leaves and minute herbaceous hermaphrodite or unisexual flowers. Stamens when equalling lobes of perianth in number, opposite to them. Ovary superior, one-celled, with a solitary ovule. Seed albuminous, with a curved embryo.

AMARANTACEÆ (p. 268) —Herbs or shrubs (*Deeringia*) with opposite or alternate exstipulate leaves and minute scarious hermaphrodite or unisexual flowers. Stamens usually five. Ovary superior, one-celled, with one ovule, or several ovules on a central placenta. Seeds albuminous, with a curved embryo.

POLYGONACEÆ (p. 269).—Herbs or shrubs with alternate simple leaves and sheathing stipules, and small greenish or coloured hermaphrodite or unisexual flowers. Ovary superior, with a solitary erect ovule; stigmas two or more. Seeds albuminous, with a straight or slightly curved embryo.

URTICACEÆ (p. 270).—Trees, shrubs, or herbs with usually alternate stipulate leaves and minute herbaceous unisexual flowers. Stamens as many as perianth-segments, and opposite to them. Ovary free. Fruit one-seeded.

EUPHORBIACEÆ (p. 276).—Trees, shrubs, or herbs with unisexual flowers. Ovary free, three-celled, with one or two pendulous ovules in each cell. Seeds albuminous.

ARISTOLOCHIACEÆ (p. 281).—Climbing shrubs or herbs with alternate leaves, and usually an irregular perianth, valvate in bud. Ovary inferior, three- to six-celled; ovules indefinite.

NEPENTHACEÆ (p. 281).—Climbing shrubs with alternate pitcher-bearing leaves, and racemose diœcious flowers.

SALICACEÆ (p. 282).—Trees or shrubs with alternate leaves, and diœcious amentaceous flowers. Perianth o, or rudimentary. Ovary free, one-celled; ovules indefinite, basal or parietal.

CUPULIFERÆ (p. 284).—Trees with alternate stipulate simple leaves and monœcious flowers. Ovary inferior, surmounted by a rudimentary, toothed perianth-limb ; two- or more-celled. Fruit one-celled, one-seeded. Seed exalbuminous.

THYMELACEÆ (p. 286).—Shrubs with a tenacious bark. Flowers usually hermaphrodite. Stamens definite. Ovary free, one-celled, with one pendulous ovule (except *Aquilaria*).

SANTALACEÆ (p. 287).—Herbs, shrubs, or trees with alternate or opposite entire leaves. Flowers hermaphrodite or unisexual. Stamens opposite to the perianth-lobes. Ovary inferior, with few ovules suspended from a free central placenta.

ELÆAGNACEÆ (p. 289).—Trees or shrubs more or less covered with silvery scurf-scales. Flowers usually hermaphrodite. Base of perianth-tube persistent around the free one-celled ovary. Ovule one, erect.

MYRISTICACEÆ (p. 290).—Trees or shrubs, with alternate entire leaves and inconspicuous diœcious flowers. Stamens monadelphous. Ovary free, one-celled, with one erect ovule. Seed with ruminated albumen.

LAURACEÆ (p. 291).—Trees or shrubs (or leafless parasites in *Cassytha*) with entire, usually evergreen, leaves. Flowers hermaphrodite or unisexual. Anthers opening by recurved valves. Ovary one-celled, free, with one pendulous ovule. Albumen o.

PIPERACEÆ (p. 292).—Jointed shrubs or herbs with alternate or opposite simple leaves. Flowers in spikes (or racemes), hermaphrodite or unisexual. Perianth o. Stamens two, or three. Ovary one-celled, one-ovuled. Seed albuminous.

** *Gymnospermous* (Ovules fertilised by direct contact of the pollen).

CONIFERÆ (p. 294).—Branching trees with simple, usually acicular or linear leaves.

CYCADACEÆ (p. 299).—Unbranched trees with a terminal crown of pinnate leaves.

MONOCOTYLEDONS.

PALMACEÆ (p. 300).—Stem woody, erect, or slender and scrambling, or acaulescent.   Leaves very large, in terminal tufts ; pinnately or palmately incised or compound.   Perianth six-leaved.   Ovary free, of three distinct or united carpels.

PANDANACEÆ (p. 303).—Stem woody or herbaceous. Leaves linear, sheathing (pinnate in *Nipa*).   Flowers unisexual, sessile in heads or spikes.   Perianth o (except in staminate flowers of *Nipa*).   Ovary one-celled.

TYPHACEÆ (p. 304).—Marsh herbs with linear leaves and spicate or capitate unisexual flowers.   Perianth o.   Fruit a dry, one-seeded nut.

AROIDEÆ (p. 305).—Stem herbaceous or woody, often scandent, or acaulescent. Leaves usually net-veined. Flowers unisexual or hermaphrodite, sessile on a spadix.   Perianth o, or of minute scales.

PISTIACEÆ (p. 308).—Floating herbs, in *Lemna* consisting of minute leaf-like fronds.   Spadix adnate to the spathe. Perianth o.   Ovary one-celled, with indefinite laterally affixed ovules.

TACCACEÆ (p. 309).—Herbs with radical, entire or divided leaves. Flowers umbellate, on long scapes, regular, hermaphrodite.   Perianth six-lobed.   Ovary inferior, with numerous ovules on three parietal placentas.

DIOSCOREACEÆ (p. 309).—Usually twining herbs with net-veined simple or digitate leaves. Flowers unisexual. Perianth six-lobed. Ovary inferior, three-celled.

LILIACEÆ (p. 310).—Herbs (in *Dracæna* shrubs or tree-like). Perianth six-leaved, regular, coloured. Stamens six. Ovary superior, three-celled.

JUNCACEÆ (p. 312).—Herbs with grass-like or quill-like leaves. Perianth six-leaved, regular, scarious. Stamens six. Ovary superior.

COMMELYNACEÆ (p. 313).—Herbs, the leaves usually deeply sheathing at the base. Perianth nearly regular, the three inner segments petaloid, three outer herbaceous. Stamens six, or fewer. Ovary superior, three- or two-celled.

ERIOCAULONEÆ (p. 314). — Aquatic or marsh herbs. Flowers minute, unisexual, in terminal heads. Perianth of staminate flowers tubular ; of pistillate, three-leaved.

PONTEDERIACEÆ (p. 314).—Aquatic herbs. Flowers hermaphrodite, petaloid, irregular, racemose from the sheath of the upper or only leaf of the scape. Ovary superior.

ORCHIDACEÆ (p. 315).—Epiphytal or terrestrial herbs, rarely scandent. Flowers irregular. Stamen one (except *Cyripedium* with two), consolidated with the stigma. Ovary inferior.

BURMANNIACEÆ (p. 322).—Herbs with grass-like or scaly leaves. Flowers regular, hermaphrodite. Perianth coloured. Stamens three to six. Ovary inferior.

SCITAMINEÆ (p. 322).—Herbs with irregular coloured flowers. Stamen one, free (except in *Musa*). Ovary inferior, three-celled.

AMARYLLIDEÆ (p. 326).—Herbs with regular coloured flowers. Perianth six-lobed. Stamens six. Ovary inferior, three-celled.

IRIDACEÆ (p. 329).—Herbs with regular or irregular

coloured flowers. Stamens three. Ovary inferior, three-celled.

HYDROCHARIDACEÆ (p. 329). — Submerged or floating herbs. Flowers usually unisexual. Perianth three- to six-leaved. Ovary inferior. Seeds exalbuminous.

ALISMACEÆ (p. 331).—Aquatic herbs. Perianth free, six-leaved ; three inner leaves petaloid. Pistil apocarpous. Seeds exalbuminous.

NAIADACEÆ (p. 332).—Floating or submerged herbs. Perianth minute, four-leaved or o. Stamens one, two, or four. Ovary free, of one, two, or four distinct carpels. Seed exalbuminous.

CYPERACEÆ (p. 333) —Grass-like herbs. Sheaths of the leaves entire (not split). Flowers hermaphrodite or uni-sexual, naked or perianth reduced to bristles ; each in the axil of a single scaly bract. Embryo at the base of copious albumen.

GRAMINEÆ (p. 335).—Grasses. Herbs (except in *Bambusa* and allies). Sheaths of the leaves usually split. Flowers usually perfect, sheathed by two-rowed scaly bracts (glumes), of which the innermost (pale) is usually two-nerved. Embryo obliquely applied at the base of copious albumen.

CHAPTER III.

CLASS I.—DICOTYLEDONS.

SUB-CLASS—*Dichlamydeæ.* DIVISION—*Thalamiflora.*

* *Pistil distinctly apocarpous* (excepting in some Dilleniaceæ, Nymphæaceæ, and the genus Anona).

1. Natural Order, *Ranunculaceæ.*—The Ranunculus Family.

Herbs with radical and cauline or floating leaves, or climbing shrubs with opposite leaves. Petals five, fewer, or o ; if o, the sepals are petaloid. Stamens usually indefinite. Seeds without an aril.

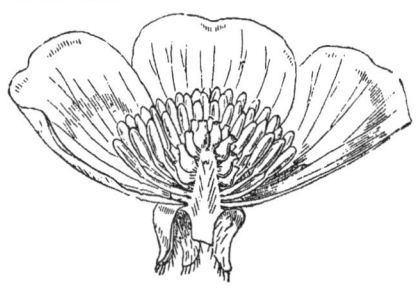

FIG. 96. Vertical section of flower of Ranunculus.

TYPE—*Ranunculus* (any species).

Herbs usually growing in damp places (one species floating in water), with entire or divided, simple, radical

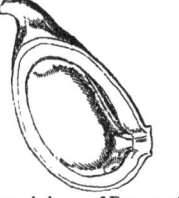

FIG. 97. Achene of Ranunculus laid open.   FIG. 98. Vertical section of seed of Ranunculus.

and cauline leaves, and regular yellow or white flowers, with deciduous sepals.

Organ.	No.	Cohesion.	Adhesion.
Calyx. *sepals.*	5(-3)	Polysepalous.	Inferior.
Corolla. *petals.*	5(-15)	Polypetalous.	Hypogynous.
Stamens.	∞	Polyandrous.	Hypogynous.
Pistil. *carpels.*	∞	Apocarpous.	Superior.
Seed solitary, albuminous. Embryo minute.			

The deviations from this type are so considerable that reference should be made to three Sub-types, each of which is represented in India as well as in Europe, though in India generally confined to the Himalaya, very few occurring in the temperate mountainous regions of the south, and still fewer in the hot plains.

SUB-TYPE 1 (*Clematis*).—Stem woody, usually climbing. Leaves opposite, compound. Sepals petaloid, valvate in bud. Petals o, or inconspicuous. Carpels ∞ .

SUB-TYPE 2 (*Anemone*).—Low herbs, with divided, simple leaves. Flowers regular, with a leafy involucre. Sepals imbricate. Petals o. Carpels ∞.

SUB-TYPE 3 (*Aconitum*).—Perennial herbs. Flowers irregular. Sepals 5, petaloid ; the upper one helmet-shaped (*galeate*). Petals 2, spurred ; the rest suppressed or very small. Carpels 3 or 6, several-seeded.

Compare the fruits of

Clematis, *Ranunculus*, or *Anemone* . . . achene.

Actæa (Himalayan plant identical with an
      English species) . . . . berry.

Aconitum (Monkshood), or *Delphinium*
      (Larkspur) . . . . follicle.

Nigella (a garden annual) . . . . . . capsule.

This Family, relatively abundant in temperate and arctic countries, does not contribute materially to Indian tropical vegetation.

Many of the species have an acrid juice, and some are dangerous poisons, as the Aconite, especially *Aconitum ferox* of the temperate Himalaya, which is one of the species affording the Bikh poison. Other species are useful in medicine ; amongst the rest, some of the Aconites and the Mishmee " Teeta " (*Coptis*), the latter a herb of Upper Assam, scarcely known to botanists, although the bitter root is collected as a drug by native tribes.

2. Natural Order, *Dilleniaceæ.*—The Dillenia Family.

Trees, shrubs, or herbs with distinctly alternate or radical leaves. Sepals persistent. Petals 5 or 4. Stamens indefinite. Seeds with an aril.

TYPE—*Dillenia speciosa.*

A spreading tree of moderate size, with alternate, simple,

serrate leaves, and large, white, solitary, regular flowers, with persistent sepals.

Organ.	No.	Cohesion.	Adhesion.
Calyx. *sepals.*	5	Polysepalous.	Inferior.
Corolla. *petals.*	5	Polypetalous.	Hypogynous.
Stamens.	∞	Polyandrous.	Hypogynous.
Pistil. *carpels.*	∞	Syncarpous.	Superior.
Seeds ∞, albuminous, hairy.			

OBSERVE the climbing genera, *Delima, Tetracera*, and *Schumacheria;* the herbaceous Ceylonese and Peninsular genus *Acrotrema;* the normally apocarpous fruit and arillate seeds of the Family.

The type-species (*Dillenia speciosa*) presents, in respect of the pistil, an aberrant character in the Order, the numerous carpels being consolidated with the fleshy axis.

Compare the apocarpous fruit, with dehiscent carpels, of *Delima, Tetracera*, and *Wormia*, with the indehiscent fruit of *Dillenia*, enclosed in the thickened, fleshy, persistent sepals.

The fruit of *Dillenia* has an agreeably acid taste, and may be used in jellies.

3. Natural Order, *Magnoliaceæ.*—The Magnolia Family.

Trees or shrubs with convolute stipules (except *Illicium*). Sepals and petals trimerous (except in *Illicium*), imbricated. Albumen uniform.

TYPE—*Michelia Champaca.*

A moderate-sized tree, common in gardens all over India,

with alternate, entire, stipulate leaves, and large, axillary, sweet-scented, yellow flowers.

Organ.	No.	Cohesion.	Adhesion.
Calyx. *sepals.*	Indistinguishable, 15(-21.)	Polysepalous.	Inferior.
Corolla. *petals.*		Polypetalous.	Hypogynous.
Stamens.	∞	Polyandrous.	Hypogynous.
Pistil. *carpels.*	∞	Apocarpous.	Superior.
Seeds solitary, albuminous.			

OBSERVE the large convolvute, deciduous stipules, which sheath the leaf-buds : the tendency to an arrangement of the sepals and petals in threes, less evident in the Champaca than in some other Magnolias : the carpels of Champaca arranged upon a long receptacle, three to four inches in length when in fruit, and raised by a short stalk above the envelopes of the flower : the fruit-carpels splitting dorsally when ripe ; the seeds pendulous, by a thread-like funicle, after dehiscence of the carpels : the carpels of *Talauma,* dehiscing by the ventral suture and falling away from the axis, leaving the ripe seeds suspended to it.

FIG. 99. Star Anise (*Illicium*). The fruit apocarpous. Carpels uniseriate, dehiscing by their ventral sutures.

Compare the axillary flowers of the Champaca with the terminal flowers of the true Magnolias.

A species of Star Anise (*Illicium*), growing in the Khasia mountains, represents a section of the Family, differing from the type in the absence of stipules, and in the carpels arranged in a single whorl instead of upon an elongated receptacle.

The flowers of the Magnolias are extremely handsome and often powerfully scented. In some species, as in the magnificent *Magnolia Campbelli* of Darjiling, they appear before the leaves. Some of the Himalayan species grow to a large size, and afford a useful timber.

4. Natural Order, *Anonaceæ.*—The Custard-Apple Family.

Trees or shrubs with alternate, entire, exstipulate leaves. Sepals and petals trimerous. Stamens indefinite. Albumen ruminate.

TYPE—The Sweet-sop, *Anona squamosa* (or Custard-Apple,[1] *A. reticulata*).

West Indian trees, cultivated very extensively through the tropics, with alternate, entire, lanceolate, exstipulate leaves, and the sepals and petals in whorls of three.

Organ.	No.	Cohesion.	Adhesion.
Calyx. *sepals.*	3	Polysepalous.	Inferior.
Corolla. *petals.*	6	Polypetalous.	Hypogynous.
Stamens.	∞	Polyandrous.	Hypogynous.
Pistil. *carpels.*	∞	Syncarpous.	Superior.
Seeds one in each cell, with ruminated albumen.			

The above species are selected as types, because they are so generally cultivated throughout India, whilst very few

[1] In India, *A. squamosa* is everywhere called the Custard-Apple; while *A. reticulata* is sometimes called the Sweet-sop.

of the native species are widely spread. Their well-known fruits, however, are not characteristic of the Family. In all other Indian species, one hundred and twenty or one hundred and thirty in number, the carpels are entirely

FIG. 100. Custard-Apple (*Anona reticulata*), flowering branch, with detached fruit (much reduced).

distinct, so that the fruit is apocarpous. In other respects the Custard-Apple is typical.

OBSERVE the woody, hooked peduncles of the Sweet-scented *Artabotrys*, common in gardens: the trimerous symmetry of the flowers exceptional amongst Dicotyledons: the valvate æstivation of the sepals and petals in most of the Family: the variable number of carpels and of seeds in each fruit-carpel in different genera; comparing the one-seeded carpels of *Guatteria longifolia* with the (usually) several-seeded moniliform carpels of *Unona discolor*, both common in gardens. Note, also, the constantly ruminated albumen of the seeds.

5. Natural Order, *Menispermaceæ.*—The Moonseed Family.

Climbing or twining shrubs with alternate leaves, and very small diœcious flowers. Stamens definite, often monadelphous. Carpels usually three.

Fig. 101. *Tinospora cordifolia;* the lower figure with young achenes (reduced).

TYPE—*Tinospora cordifolia.*

A common straggling, woody, climbing or twining plant, with corky, warted bark, alternate, petiolate, cordate leaves, and axillary racemes of small, yellow, diœcious flowers.

Organ.	*No.*	*Cohesion.*	*Adhesion.*
Calyx. *sepals.*	6	Polysepalous.	Inferior.
Corolla. *petals.*	6	Polypetalous.	Hypogynous.
♂ Stamens.	6	Hexandrous.	Hypogynous.
♀ Pistil. *carpels.*	3	Apocarpous.	Superior.
Seeds solitary, curved, with ruminated albumen.			

OBSERVE the broad medullary-rays in a cross-section of the stem of any of the larger species, as *Cissampelos Pareira :* the formation of successive concentric rings of distinct vascular bundles, also well shown in cross-sections of thick-stemmed species : the development of thread-like adventitious roots from the stem of *Tinospora*, sometimes thirty feet long: the flowers of the Family, usually of trimerous symmetry, but always inconspicuous and unisexual: the two sepals united into a single scale in the female flower, and the four petals united into a minute cupshaped corolla in *Cissampelos Pareira :* the stamens opposite to the petals : the sterile stamens in the female flower of the type : the characteristic horse-shoe shape of the seed.

The Family is chiefly tropical, and nearly all the species are climbers. *Cocculus laurifolius* of the Himalaya is exceptional in this respect, forming a small erect tree or shrub.

The berries of *Anamirta* (*Cocculus indicus*) are poisonous, and are used to kill fish in India. The tonic medicinal Calumba-root is afforded by an African species of *Jateorhiza* (*J. Columba*), indigenous on the Mozambique coast. The root of the common *Tinospora* (Gulancha) is also used in medicine.

6. Natural Order, *Nymphæaceæ.*—The Water-Lily Family.
    Aquatic herbs.   Petals and stamens indefinite.
    TYPE—Lotus Water-lily (*Nymphæa Lotus*).

Organ.	No.	Cohesion.	Adhesion.
Calyx. *sepals.*	4	Polysepalous.	Inferior.
Corolla. *petals.*	∞	Polypetalous.	Outer hypogynous. Inner epigynous.
Stamens.	∞	Polyandrous.	Peri- or epi-gynous.
Pistil. *carpels.*	∞	Spuriously syncarpous.	Superior.
Seeds ∞ albuminous ; albumen double.			

Representing distinct Sub-types are :—

The Sacred Lotus (*Nelumbium speciosum*), Padma or Pudma of the Hindoos, in which the carpels are separately immersed in a top-shaped receptacle, and the seeds exalbuminous ; and

Brasenia, with small flowers, three (petaloid) sepals, three petals, and free carpels.

OBSERVE the wide air-cavities in the petioles and peduncles : the very gradual transition from sepals to petals, and from petals to stamens, in the flowers   the receptacle which developes around and adnate to the carpels, so that they appear united into a syncarpous pistil, and the petals and stamens seem as though inserted upon the ovary: the arrangement of the ovules, which are spread over the sides of the ovaries : the double albumen of the seeds of the true

Water-lilies ; the inner albumen, next to the embryo, being
formed within the embryo-sac (*endosperm*), the outer albumen
the remains of the tissue of the nucleus of the ovule
(*perisperm*).

FIG. 102. Sacred Lotus (*Nelumbium speciosum*). About one-tenth to
one-fifteenth natural size.

Victoria regia, a South American member of the Family,
nearly allied to the prickly *Euryale* of India, bears floating
leaves which have been measured twelve feet across, and
flowers about one foot in diameter when expanded.

Besides the points noted above, *Nelumbium* differs from
the true Water-lilies in its peduncles and petioles rising

high above the surface of the water. The leaves are peltate, and sometimes one to two feet in diameter. The rhizome and seeds are eaten, as are those of the Nymphæas.

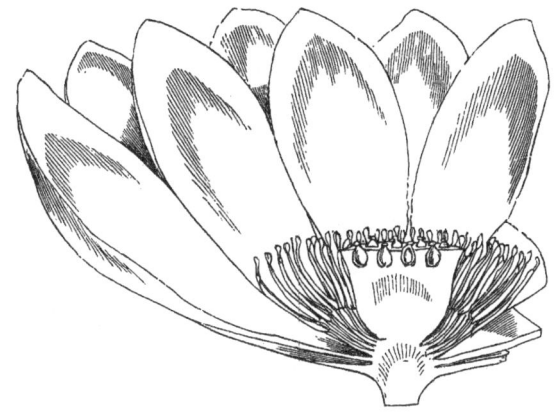

FIG. 103. Vertical section of flower of *Nelumbium*, showing hypogynous stamens, and carpels singly immersed in a turbinate receptacle.

** *Pistil syncarpous* (except Connaraceæ and some Anacardiaceæ).

7. Natural Order, *Papaveraceæ.*—The Poppy Family.

Herbs with milky, coloured, or (Tribe *Fumarieæ*) watery juice.   Stamens indefinite, free, or (*Fumarieæ*) definite, diadelphous.   Ovules parietal.   Seeds albuminous.

TYPE—Opium Poppy (*Papaver somniferum.*)

An annual, erect herb, with milky juice, and large, solitary, terminal, white or purple fugacious flowers.

FIG. 104. Opium Poppy (*Papaver somniferum*), with detached capsule.   One-third
to one-fourth natural size.

Organ.	No.	Cohesion.	Adhesion.
Calyx. *sepals.*	2	Polysepalous.	Inferior.
Corolla. *petals.*	4	Polypetalous.	Hypogynous.
Stamens.	∞	Polyandrous.	Hypogynous.
Pistil. *carpels.*	∞	Syncarpous.	Superior.
Seeds ∞, minute, with oily albumen.			

OBSERVE the caducous sepals, thrown off as the crumpled
petals expand : the tendency to multiplication of the petals
at the expense of the stamens in this species and in the

small Scarlet Poppy (*P. Rhœas*) when grown in gardens : the partial dissepiments of the ovary (opposite to the lobes of the stigma) which are not coherent in the middle : the placentary or ovule-bearing surface, spread over the sides of the projecting plates : the dehiscence of the capsule by pores around the top.

The yellow-flowered Mexican Poppy (*Argemone Mexicana*), with a capsule dehiscing by short valves, is an introduced and very common weed in India. The Opium Poppy is largely cultivated in the north, and occurs as a weed in waste places. The drug opium is the inspissated, milky juice collected from punctures in the unripe capsules. The seeds afford a valuable oil.

8. Natural Order, *Cruciferæ.*—The Crucifer Family.

Herbs with alternate, exstipulate leaves.   Sepals four. Petals four.   Stamens usually tetradynamous.   Seeds exalbuminous.

TYPE—Mustard or Rape ; species of *Brassica* (either will serve).

Herbs with alternate, more or less lobed or pinnatifid exstipulate leaves, and terminal racemes of ebracteate, cruciform, yellow flowers.

Organ.	No.	Cohesion.	Adhesion.
Calyx. *sepals.*	4	Polysepalous.	Inferior.
Corolla. *petals.*	4	Polypetalous.	Hypogynous.
Stamens.	6	Tetradynamous.	Hypogynous.
Pistil. *carpels.*	2	Syncarpous.	Superior.
Seeds exalbuminous, cotyledons folded on radicle.			

This Family, very abundant and of great importance in cool climates, has but few native representatives in India. I have selected two species cultivated in the cool season for the sake of the oil contained in their seeds, and which are probably more easily obtainable than native species.

FIG. 105. Indian Mustard (*Brassica juncea*). Lower part of stem and inflorescence about one-half natural size. A detached siliqua to the right.

The Turnip, Radish, Cress, Seakale, and Cabbage of temperate countries, all belong to this Family.

OBSERVE the six stamens, of which two are shorter than the rest (tetradynamous), explained by assuming that they belong to two whorls, two stamens of the outer whorl being suppressed : the two-valved fruit (*siliqua*) dehiscing from below, upwards. In Radish (*Raphanus*) the siliqua is indehiscent, and in Brassica it is often three-celled.

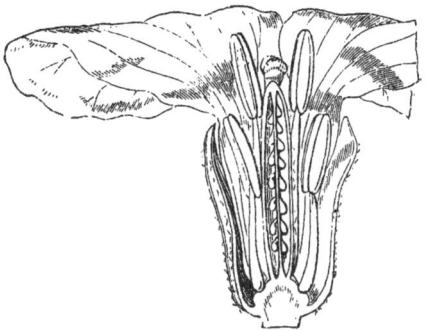

Fig. 106. Vertical section of flower of a Crucifer, enlarged.

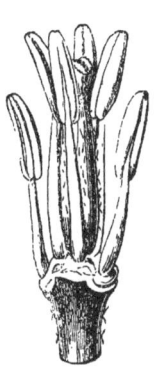

Fig. 107. Tetradynamous stamens and pistil of a Crucifer.

Fig. 108. Capsule (siliqua) of a Crucifer ; the valves separating from below upwards.

9. Natural Order, *Capparidaceæ.*—The Caper Family.

Herbs or shrubs with alternate, often divided leaves.

M

Sepals 4. Petals 4. Stamens indefinite, 8, 6 (not tetra-dynamous), or 4. Ovary often stipitate. Seeds usually exalbuminous.

TYPE—*Gynandropsis pentaphylla.*

An erect, hairy annual, with alternate, quinque-foliolate leaves, and terminal, bracteate racemes, of small white or pale pink flowers.

Organ.	No.	Cohesion.	Adhesion.
Calyx. *sepals.*	4	Polysepalous.	Inferior.
Corolla. *petals.*	4	Polypetalous.	Hypogynous.
Stamens.	6	Hexandrous.	Hypogynous.
Pistil. *carpels.*	2	Syncarpous.	Superior.
Seeds ∞, exalbuminous, embryo curved.			

OBSERVE the long stalk (*gynophore*) supporting the stamens and ovary above the envelopes of the flower. Compare the siliquiform fruit with the more or less berry-like, often indehiscent fruit of *Cadaba, Capparis* (true Capers), and *Cratæva.*

In *Polanisia viscosa*, a common Indian weed with yellow flowers, the ovary is sessile.

The Family is closely allied to the Crucifers, differing from them chiefly in habit and in their stamens, which are usually more numerous, and the ovary raised upon a stalk (*gynophore*). Our type-species is a common weed of cultivated land in India, where the leaves are used in curries.

10. Natural Order, *Violaceæ.*—The Violet Family.

Trees, shrubs, or herbs with alternate stipulate leaves. Sepals 5. Petals 5. Stamens 5, connective usually dilated or prolonged above the anther-cells. Ovary one-celled ; placentas three.

TYPE—Any species of Violet or Pansy (*Viola*).

Low herbs, with alternate or radical, stipulate, simple leaves, and axillary, solitary, irregular flowers.

Organ.	No.	Cohesion.	Adhesion.
Calyx. sepals.	5	Polysepalous.	Inferior.
Corolla. petals.	5	Polypetalous.	Hypogynous.
Stamens.	5	Pentandrous.	Hypogynous.
Pistil. carpels.	3	Syncarpous.	Superior.
Seeds parietal, albuminous.			

OBSERVE the " spur" of the lower petal, technically the upper one, though the lower in position, owing to the flower being inverted (*resupinate*) : the small, almost apetalous, closed flowers of some species, appearing after the large flowers are past ; they are self-fertilized, and bear fruit with numerous seeds : the dehiscence of the fruit in three valves by the dorsal sutures of the carpels, so that each valve bears a parietal row of seeds.

FIG. 109. Transverse section of ovary of Viola, showing three parietal multiovulate placentas.

A marked Sub-type of this Family (*Alsodeia*) occurs in the Khasia hills,

M 2

Ceylon, and in the Malay Peninsula, differing in arborescent habit, and small, nearly regular flowers.

11. Natural Order, *Bixaceæ.*—The Annatto Family.

Trees or shrubs with alternate, simple leaves. Flowers unisexual or hermaphrodite, often apetalous. Stamens indefinite. Seeds albuminous.

TYPE—The Annatto (*Bixa Orellana*).

A small tree, with alternate, entire leaves, terminal panicles of beautiful white flowers and echinate capsules. (A South American tree cultivated in India.)

Organ.	*No.*	*Cohesion.*	*Adhesion.*
Calyx. *sepals.*	5	Polysepalous.	Inferior.
Corolla. *petals.*	5	Polypetalous.	Hypogynous.
Stamens.	∞	Polyandrous.	Hypogynous.
Pistil. *carpels.*	2	Syncarpous.	Superior.
Seeds ∞, albuminous, coated with red pulp.			

Two distinct Sub-types occur in India, represented by—

1. *Flacourtia sepiaria*, a small spinose shrub, with simple, serrate leaves, and unisexual, apetalous flowers, bearing pleasantly acid fruit, common on uncultivated land ; and

2. *Hydnocarpus inebrians*, a large tree, with alternate, serrate leaves, and diœcious, axillary, white flowers.

OBSERVE the loculicidal dehiscence of the capsules of *Bixa*, and the red pulp of the seed, which is collected in South America for export to Europe as Annatto, used as a silk-dye and to stain cheese.

12. Natural Order, *Polygalaceæ.*—The Milkwort Family.

Trees, shrubs, or herbs with alternate, simple leaves. Sepals unequal. Stamens usually eight, often more or less coherent and also adherent to the petals.

TYPE—Polygala arvensis.

A low, procumbent, branching herb, with small entire leaves, and racemose, irregular, inconspicuous flowers.

Organ.	*No.*	*Cohesion.*	*Adhesion.*
Calyx. *sepals.*	5	Polysepalous.	Inferior.
Corolla. *petals.*	3	Polypetalous.	Epistaminal.
Stamens.	8	Monadelphous.	Hypogynous.
Pistil. *carpels.*	2	Syncarpous.	Superior.
Seeds hairy, one in each cell of the fruit.			

OBSERVE the three outer sepals, which are very small ; the two innermost, called *wings*, much larger, resembling bracts or petals, and enlarging after flowering, so as to enclose the fruit : the cohesion of the filaments and the adhesion of the really distinct petals to their tube : the small bearded crest at the end of the keel-petal : the one-celled anthers : the hooded stigma covering the anthers : the cap-like arilloid covering at the top of the seeds.

In *Salamonia* the sepals are nearly equal, and the stamens fewer.

Xanthophyllum is an arborescent Indian genus, with five petals, four of which are nearly equal, and eight stamens more or less free, with two-celled anthers.

13. Natural Order, *Caryophyllaceæ.*—The Pink Family.

Herbs with opposite, simple leaves.   Ovary one-celled, with a free central placenta.

TYPE—Indian Pink (*Dianthus chinensis*).

A garden herb, with opposite, grass-like, glabrous leaves, and terminal, solitary or fascicled red flowers blotched with purple.

Organ.	No.	Cohesion.	Adhesion.
Calyx. *sepals.*	5	Gamosepalous.	Inferior.
Corolla. *petals.*	5	Polypetalous.	Hypogynous.
Stamens.	10	Decandrous.	Hypogynous.
Pistil. *carpels.*	2	Syncarpous.	Superior.
Seeds ∞, upon a free central placenta.			

FIG. 110. Transverse section of ovary, showing free central placentation.

OBSERVE the central placenta, free from the walls of the ovary, by the early rupture or suppression of the dissepiments.

Species of this Family abound in the temperate zone of the northern hemisphere, but are rare in the tropics, excepting some small-flowered weed-like plants and a few grown in gardens.

The eastern species, taken as a type, is commonly cultivated in Indian gardens.

Low-tufted species of this Family grow at a great elevation in the Himalayas, one attaining a height of 14,000 to 18,000 feet.

14. Natural Order, *Hypericineæ.*—The St. John's Wort Family.

Herbs, shrubs, or rarely trees, with opposite, undivided leaves. Flowers hermaphrodite. Calyx imbricate.

TYPE—*Hypericum japonicum.*

A low, procumbent or ascending, perennial, glabrous herb, with opposite, entire, dotted leaves, and small, yellow, terminal, cymose flowers.

Organ.	No.	Cohesion.	Adhesion.
Calyx. *sepals.*	5	Polysepalous.	Inferior.
Corolla. *petals.*	5	Polypetalous.	Hypogynous.
Stamens.	∞	Polyadelphous.	Hypogynous.
Pistil. *carpels.*	3	Syncarpous.	Superior.
Seeds ∞, exalbuminous.			

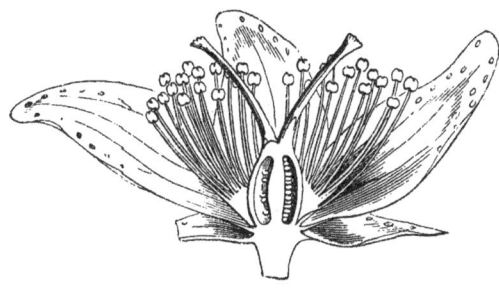

FIG. 111. Vertical section of flower of a species of Hypericum.

OBSERVE the immersed glands of the leaves, shown as

translucent dots when held up to the light : also the black, glandular dots of the sepals and petals.

The flowers of *H. mysorense* of South India are much larger than those of the type-species.

15. Natural Order, *Guttiferæ.*—The Gamboge Family.

Trees or shrubs with opposite, undivided leaves. Flowers unisexual or polygamous. Calyx imbricate.

Type—*Calophyllum Inophyllum.*

A tree with opposite, simple, coriaceous, shining, closely veined, entire leaves, and axillary, drooping racemes of fragrant, white, polygamous flowers.

Organ.	No.	Cohesion.	Adhesion.
Calyx. *sepals.*	4	Polysepalous.	Inferior.
Corolla. *petals.*	4	Polypetalous.	Hypogynous.
♂ Stamens.	∞	Polyadelphous.	Hypogynous.
♀ Pistil. *carpels.*	?	Syncarpous.	Superior.
Seeds solitary, exalbuminous.			

The pistil being normally syncarpous in this Family, it is thus described above, although the ovary is but one-celled in *Calophyllum.* Compare with the embryo of the type-species, which has large cotyledons and a short radicle, that of the Mangosteen, grown in the Straits of Malacca,.or of any other *Garcinia* or *Xanthochymus.* In the latter genera the embryo consists of an enormous radicle (*tigellum* strictly), the cotyledons being obsolete.

This well-marked tropical Family is represented in India

by six genera, including the genus *Garcinia*, which, besides the Mangosteen, includes the species affording the drug and pigment gamboge. The best gamboge is a gum exuded from wounds in the bark of a *Garcinia* growing in Siam and the Malay Peninsula, closely allied to, if not identical with, a Ceylon species. The Family generally is characterised by a coloured, resinous juice.

The flowers are usually diœcious or polygamous, with the sepals and petals in fours.

Pinnay-oil is obtained from the seeds of our type-species by pressure. It is used as a lamp-oil and in medicine.

16. Natural Order, *Ternströmiaceæ.*—The Tea Family.

Trees or shrubs, usually with alternate, simple leaves. Flowers hermaphrodite or unisexual. Calyx imbricate.

TYPE—The Tea-shrub (*Thea chinensis*).

An evergreen shrub, with alternate, shining, simple leaves, and axillary, pedunculate, white flowers.

Organ.	No.	Cohesion.	Adhesion.
Calyx. *sepals.*	5	Polysepalous.	Inferior.
Corolla. *petals.*	5	Gamopetalous.	Hypogynous.
Stamens.	∞	Polyandrous.	Hypogynous and epipetalous.
Pistil. *carpels.*	3	Syncarpous.	Superior.
Seeds solitary, exalbuminous.			

OBSERVE the slight cohesion of the petals, which character is not infrequent in several normally polypetalous Families, and is, alone, by no means of sufficient absolute importance to warrant such a genus as Tea being classed with corolli-floral (*gamopetalous*) families. Note, also, the stamens, the innermost of which are hypogynous, while the outer ones are more or less monadelphous and adherent to the base of the petals.

Compare the woody capsule, dehiscing loculicidally, of *Thea* and *Camellia* (scarcely generically distinct) with the indehiscent berries of *Saurauja* and *Eurya*.

The Tea-shrub, of recent years cultivated with so much success on the cool slopes of the mountains of Northern India, is found wild in the jungles of Assam. It has not been noticed in the wild state in China, where its cultivation dates from an extremely remote period.

17. Natural Order, *Dipterocarpeæ.*—The Malay-Camphor Family.

Trees or climbing shrubs, usually resinous, with alternate, penni-veined, simple leaves. Two or more of the segments of the calyx-limb usually enlarged in fruit.

Type—The Sal (*Shorea robusta*).

Organ.	No.	Cohesion.	Adhesion.
Calyx. sepals.	5	Gamosepalous.	Inferior.
Corolla. petals.	5	Polypetalous (?).	Hypogynous.
Stamens.	∞	Polyandrous.	Hypogynous.
Pistil. carpels.	3	Syncarpous.	Superior.
Seeds exalbuminous.			

A gigantic timber-tree, with alternate, entire, parallel-veined leaves, loose terminal and axillary panicles of yellow, hermaphrodite, regular flowers, and enlarging (accrescent) calyx-lobes.

FIG. 112. Sal (*Shorea robusta*), about one-half or one-third natural size.

OBSERVE the remarkable enlargement of each of the lobes of the calyx after flowering   in *Dipterocarpus* and *Hopea* only two become enlarged: hence the name of the former genus and of the Family, significant of the two-winged fruit. In some species of *Dipterocarpus* the wings attain several inches in length.

The Family abounds in a resin which is collected from
several species, as *Vateria indica*, which affords Indian
Copal.   Our type-species, besides furnishing a balsamic

FIG. 113. Fruit of a species of *Dipterocarpus :* the calyx persistent ; the tube
enclosing the ovary and two of the lobes accrescent.

resin, yields one of the most valuable of Indian timbers.
Sumatra Camphor is the resin of a *Dryobalanops*.   It is

eagerly bought by the Chinese, who export ordinary Camphor, their own produce, to Europe.

18. Natural Order, *Malvaceæ.*—The Mallow Family.

Shrubs, trees, or herbs with alternate, simple leaves. Calyx valvate. Stamens monadelphous, with one-celled anthers.

Type—Rose Hibiscus (*Hibiscus rosa-sinensis*).

A large garden shrub, with alternate, simple, serrate, shining leaves, and large, showy (red, white, or yellow) axillary flowers.

Fig. 114. Vertical section of flower of Rose Hibiscus, nearly of natural size. To the left a one-celled anther and free extremity of filament.

Organ.	No.	Cohesion.	Adhesion.
Calyx. *sepals.*	5	Gamosepalous.	Inferior.
Corolla. *petals.*	5	Polypetalous.	Inserted on stamens.
Stamens.	∞	Monadelphous.	Hypogynous.
Pistil. *carpels.*	5	Syncarpous.	Superior.

OBSERVE the whorl of numerous, narrow bracts, immediately under the calyx ; in Cotton (*Gossypium*) the bracts
are three in number, very large and cordate, and must not
be mistaken for sepals : the valvate æstivation of the calyx-
lobes and imbricate (contorted) æstivation of the petals :
the staminal tube five-toothed at the top, giving off, from the
side of the tube, very numerous filaments, bearing one-
celled anthers : the large-grained pollen, which, examined
under a magnifier, is found to be rough with minute
projecting points.

Plants of this Family are destitute of noxious properties.
Many of them are mucilaginous, and the liber affords a
useful fibre. The species of pre-eminent importance are the
cotton-producing plants belonging to the genus *Gossypium*,
several varieties of two or three species of which are now
very extensively cultivated in India. Cotton consists of the
delicate, long, thin-walled hairs which clothe the seeds.
These hairs, when dry, become flattened and twisted. The
commercial value of the Cotton depends upon the length
and tenacity of these hair-cells. A useful lamp-oil is
expressed from the seeds of the Cotton-plant, and the
refuse may be compressed into an oil-cake for feeding
cattle.

The Okra (*Hibiscus esculentus*), a tropical American species, is much cultivated in India for the sake of its unripe mucilaginous capsules, which are used as an article of diet.

FIG. 115. Vertical section of flower of Mallow (*Malva*), slightly enlarged.

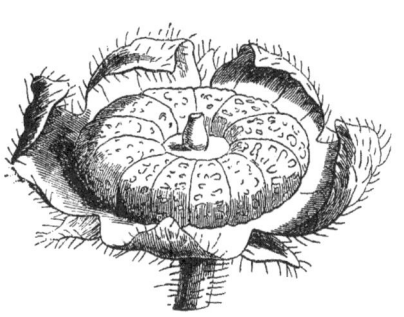

FIG. 117. Embryo of Mallow with folded cotyledons.

FIG. 116. Fruit of Mallow. The carpels uniseriate.

The Durian fruit is the produce of a Malayan tree of this Family (*Durio zibethinus*), to which belong, also, the African Baobab (*Adansonia*) and the so-called Cotton-tree

(*Bombax*).   These latter agree in the general structure of their flowers with the type, but they are large trees, and the two last-named have digitate leaves.

19. Natural Order, *Sterculiaceæ.*—The Sterculia Family.

Trees, shrubs, or herbs with alternate simple or digitate leaves.   Calyx valvate.   Stamens monadelphous, with two-celled anthers.

TYPE—*Sterculia fœtida.*

A large, deciduous tree, with alternate digitate leaves crowded towards the ends of the branches, and racemose, crimson and brown, hairy, unisexual flowers.

Organ.	No.	Cohesion.	Adhesion.
Calyx. *sepals.*	5	Gamosepalous.	Inferior.
Corolla.	o	...	...
♂ Stamens.	∞	Monadelphous.	Hypogynous.
♀ Pistil. *carpels.*	5	Syncarpous.	Superior.

OBSERVE the two-celled anthers, collected into a small, close head : in the female flower a few imperfect stamens may be found immediately under the stipitate ovary. Compare with the type, *Helicteres* (*Isora*), and several other common Indian genera of the Family (*Pterospermum, Pentapetes, Waltheria*, &c.), in which the corolla is present and the flowers are hermaphrodite.   Observe the unequal petals of *Helicteres* and the remarkable spirally-twisted carpels of the fruit.

The Chocolate and Cocoa-tree (*Theobroma*), cultivated in the northern provinces of South America, Central America.

and some of the West Indian Islands, belongs to this family.   Cocoa and Chocolate are prepared from the seeds, which are roasted, ground, and usually mixed with sugar, arrowroot, and spices.   The seeds are closely packed in a capsule four to six inches long by two or three inches in diameter.

20. Natural Order, *Tiliaceæ.*--The Jute and Lime-tree Family.

Trees, shrubs, or herbs, usually with alternate, simple leaves, valvate calyx, and indefinite free stamens.

TYPE—Common Jute (*Corchorus capsularis*).

A smooth, annual herb, rather woody below, with alternate, simple leaves, and axillary fascicles of small, yellow, regular flowers.

Organ.	No.	Cohesion.	Adhesion.
Calyx. *sepals.*	5	Polysepalous.	Inferior.
Corolla. *petals.*	5	Polypetalous.	Hypogynous.
Stamens.	∞	Polyandrous.	Hypogynous.
Pistil. *carpels.*	5	Syncarpous.	Superior.

OBSERVE the tough fibrous liber of the bark, used to make "gunny" bags.   Compare the fruit, a roundish capsule, with the siliquæform capsule of *Corchorus olitorius*, in which the seeds are separated from each other by transverse partitions: also with the indehiscent, drupe-like, often lobed fruit of the Grewias, and the bristly capsule of the Triumfettas.   *Corchorus olitorius* is used as a pot-herb.

N

Like the Mallow and Sterculia Families, to which this Family is closely allied, the species are generally harmless, and more or less mucilaginous ; many of them have also a tenacious liber. From the bark of the European Lime-tree (*Tilia*) Russia-matting or "bast" is obtained.

The genus *Elæocarpus*, including some large Indian trees, belongs to a section of this Family, marked by the anthers dehiscing at their tips. Elæocarpus itself usually has its petals with a deeply-cut margin. The stones of its drupaceous fruits are often ornamented and strung together as beads.

21. Natural Order, *Linaceæ.*—The Flax Family.

Shrubs, herbs, or rarely trees, with alternate, undivided, simple leaves. Sepals imbricate. Stamens more or less coherent below. Styles usually free.

Type—*Linum trigynum.*

A shrubby plant, with alternate, simple, entire, smooth leaves, and axillary, regular, yellow, pretty flowers, collected near the ends of the branches.

Organ.	No.	Cohesion.	Adhesion.
Calyx. sepals.	5	Polysepalous.	Inferior.
Corolla. petals.	5	Polypetalous.	Hypogynous.
Stamens.	5	Monadelphous.	Hypogynous.
Pistil. carpels.	3	Syncarpous.	Superior.

OBSERVE the tenacity of the liber, especially in the Common Flax (*L. usitatissimum*), a cultivated, slender herb, with fugacious blue flowers : the slight cohesion of the filaments, and the minute teeth projecting in the intervals between the filaments indicating a second series of undeveloped stamens. In *Erythroxylon*, an allied genus, there are ten perfect stamens without any rudimentary ones. Observe, also, the spuriously six-celled ovary of the type, resulting from each of the three cells becoming divided into two by the infolding of the dorsal suture of each carpel. Compare, under the microscope, the fibre of Flax with Cotton. Flax consists of long, thick-walled liber-cells, resembling jointed cylindrical rods. Fibres of linen from mummy-cloth of Egypt may also be compared with cloth from the Peruvian tombs, which is made of cotton.

In the genus Flax, as in several others belonging to widely different Families, a dimorphous condition of the essential organs has been observed, consisting in differing relative lengths and position of the anthers and stigmas in different flowers. The object of this dimorphism Mr. Darwin has shown to be to secure fertilisation by the pollen of distinct flowers of the same species (see Journal of the Linnean Society of London, Botany, vol. vii. p. 69).

22. Natural Order, *Malpighiaceæ.*—The Malpighia Family.

Climbing shrubs with opposite, entire leaves. Stamens ten. Ovary three-lobed. Carpels winged when ripe.

TYPE—*Hiptage Madablota.*

Organ.	No.	Cohesion.	Adhesion.
Calyx. *sepals.*	5	Polysepalous.	Inferior.
Corolla. *petals.*	5	Polypetalous.	Hypogynous.
Stamens.	10	Decandrous.	Hypogynous.
Pistil. *carpels.*	3	Syncarpous.	Superior.

A climbing shrub with opposite, entire leaves, and irregular, very fragrant, yellow-white flowers.

OBSERVE the closely adpressed hairs, covering the young parts of the plant; examined minutely, they will be found to be attached near the middle (peltate hairs) : the single " gland " under the flower, partly adnate to the pedicel, partly to the calyx; in some genera each sepal bears one or two dorsal glands : the clawed, unequal, fringed petals : the declinate stamens : the winged fruit-carpels.

The head-quarters of the Family is in South America, where many " lianes " of the moist forests belong to it.

Very few species are turned to practical account.

23. Natural Order, *Zygophyllaceæ.*—Bean-Caper Family.

Herbs or low shrubs with opposite, stipulate, compound leaves. Peduncles axillary, one-flowered. Stamens eight or ten, free.

TYPE—*Tribulus terrestris.*

A prostrate herb with opposite, stipulate, pinnate leaves, axillary, solitary, regular flowers, and spinose fruits.

Organ.	No.	Cohesion.	Adhesion.
Calyx. *sepals.*	5	Polysepalous.	Inferior.
Corolla. *petals.*	5	Polypetalous.	Hypogynous.
Stamens.	10	Decandrous.	Hypogynous.
Pistil. *carpels.*	5	Syncarpous.	Superior.

To this Family belong also *Fagonia*, with trifoliolate leaves, common in some of the dry parts of India, and the West Indian *Guiacum ;* the latter affording the hard, heavy, greenish-brown wood called *lignum-vitæ.*

24. Natural Order, *Geraniaceæ.*—The Geranium, Sorrel, and Balsam Family.

Herbs with opposite or alternate, simple or compound leaves. Flowers regular or irregular, usually two or more on axillary peduncles.

TYPE—Any garden Geranium (*Pelargonium*)

Usually glandular-pubescent herbs, with opposite, lobed, palmi-veined leaves, and umbellate, slightly irregular, handsome flowers.

Organ.	*No.*	*Cohesion.*	*Adhesion.*
Calyx. *sepals.*	5	Polysepalous.	Inferior.
Corolla. *petals.*	5	Polypetalous.	Hypogynous.
Stamens.	7(-10)	Monadelphous.	Hypogynous.
Pistil. *carpels.*	5	Syncarpous.	Superior.

As Sub-types may be taken :—

1. Indian Cress (*Tropæolum majus*), a climbing or prostrate, succulent, garden herb, with alternate, peltate leaves, and irregular, orange-coloured, octandrous flowers, with the posterior sepal spurred.

2. Sorrel (*Oxalis*), any species. Herbs with trifoliolate or pinnate leaves, and regular, decandrous flowers.

3. Balsam (*Impatiens*), any species. Succulent or tender herbs, with alternate or opposite simple leaves, and very irregular flowers, usually with three coloured sepals, of which the posterior is spurred, three petals,* and five stamens.

---

* Two are bifid, indicating that we have really five petals, of which four cohere in pairs.

OBSERVE the spur of the posterior sepal of cultivated *Pelargonium*, which is adherent to the pedicel. Its cavity may be easily seen on cutting the pedicel across just below the calyx. In *Tropæolum* the corresponding spur is free. Note also the beak-like prolongation of the receptacle of the Geranium Tribe after flowering: the closely connivent or coherent anthers of the Balsams: the elastic valves of the fruit-capsules of the latter.

The Balsams are a pre-eminently Indian group, but the distribution of the species is very circumscribed, many of them being limited to small areas. The beautiful aquatic *Hydrocera triflora* (*Impatiens natans* of the older botanists), differing from *Impatiens* in its distinct petals and berried fruit, is, however, widely dispersed throughout India in ponds and freshwater ditches. The floating and submerged branches of this species are often several yards in length.

Some species of *Oxalis* exhibit the phenomenon of "irritability" in their leaves. The Blimbing and Carambola are the acid fruits of species of *Averrhoa*, an arborescent genus allied to *Oxalis*, native in India.

25. Natural Order, *Rutaceæ.*—The Rue and Orange Family.

Usually trees or shrubs with alternate or opposite, compound (pinnate tri- or uni-foliolate) leaves, dotted with translucent glands. Stamens as many or twice as many as petals (in *Citrus* and *Ægle* indefinite).

TYPE—The Orange-tree (*Citrus Aurantium*).

An evergreen, often spinose tree, with alternate, entire, shining, glandular-dotted leaves, and axillary, fragrant, white flowers.

FIG. 118. Orange (*Citrus Aurantium*), reduced.

Organ.	No.	Cohesion.	Adhesion.
Calyx. *sepals.*	5	Gamosepalous.	Inferior.
Corolla. *petals.*	5	Polypetalous.	Hypogynous.
Stamens.	∞	Polyadelphous.	Hypogynous.
Pistil. *carpels.*	∞	Syncarpous.	Superior.
Seeds exalbuminous ; often polyembryonous.			

OBSERVE the articulation between the petiole and the blade of the leaf, indicating that the leaf is of the compound type.

It is unifoliolate in the Orange, as in the Barberry. In other Indian genera allied to the Orange the leaves are pinnate or trifoliolate. Observe, also, the frequently broad wings on each side of the petiole : the translucent dots of the leaves, easily seen when they are held up to the light : also the glandular dots of the flowers ; similar immersed "glands," or receptacles of essential oil, are common to most of the genera of the Family : the cup-shaped disk between the stamens and ovary : the frequent occurrence of two or more exalbuminous embryos in the seeds ; they are found irregularly and closely compressed together, so that the cotyledons, normally plano-convex, are much distorted.

By examining a series of ovaries in different stages, intermediate between the flower and fruit, the development of the pulp, which ultimately fills the numerous cells of the fruit, may be observed. It originates in cellular, papillæform projections from the inside of the outer wall of the ovary.

The Tribe (*Aurantieæ*) of this large Natural Order, to which our type belongs, is characterised by a succulent, indehiscent fruit ; but in other Tribes, including by far the larger number of species, the fruit is dry, often separating, when ripe, into nuts or *cocci*.

In Common Rue (*Ruta*) the capsule is deeply four- or five-lobed. In the climbing shrub *Zanthoxylum alatum* the fruit-carpels are free and dehiscent. The flowers of the last-named are unisexual.

Besides the Orange and its congeners the Lemon, Lime, Shaddock, and Citron, the Family includes the Bael (*Ægle Marmelos*) and the Wampi (*Clausena Wampi*) : the former is in great repute as a medicine in Southern India.

26. Natural Order, *Ochnaceæ.*—The Ochna Family.

Shrubs or trees with alternate, shining, coriaceous, simple, eglandular leaves.   Ovary usually deeply lobed.   Fruit of three, five, or more distinct drupes.

Type—Ochna squarrosa.

A small, smooth tree, common in gardens, with alternate, entire leaves, and fragrant, showy, yellow flowers, in short, racemose panicles.

Organ.	No.	Cohesion.	Adhesion.
Calyx. *sepals.*	5	Polysepalous.	Inferior.
Corolla. *petals.*	5(-12)	Polypetalous.	Hypogynous.
Stamens.	∞	Polyandrous.	Hypogynous.
Pistil. *carpels.*	5(-12)	Syncarpous.	Superior.

Observe the deep divisions of the ovary between the carpels, which, as they mature, become free and drupaceous.

27. Natural Order, *Burseraceæ.*—The Myrrh Family.

Trees with resinous juice, alternate compound leaves, and small panicled or racemose flowers.   Stamens free, as many or twice as many as the petals.

Type—Boswellia thurifera.

A large timber-tree, with alternate, imparipinnate leaves, and simple axillary racemes of small, whitish flowers, clustered towards the extremities of the branches.

Organ.	No.	Cohesion.	Adhesion.
Calyx. *sepals.*	5	Gamosepalous.	Inferior.
Corolla. *petals.*	5	Polypetalous.	Hypogynous.
Stamens.	10	Decandrous.	Hypogynous.
Pistil. *carpels.*	3	Syncarpous.	Superior.

OBSERVE the yellowish or red resin which exudes from the bark. Similar resinous products are characteristic of several species of the Family ; some of which, growing in the dry regions of Western Asia, the Red Sea, and Eastern Africa, afford Myrrh, Olibanum, and so-called Balm of Gilead.

We much want authentic information as to the source of some of these fragrant resins, and travellers in the countries where they grow would do well to secure specimens of resins, and of the trees yielding them in flower and fruit, in order that they may be determined by some competent botanist.

28. Natural Order, *Meliaceæ.*—Melia Family.

Trees or shrubs with alternate, compound leaves. Flowers small, panicled. Stamens monadelphous (except in *Cedrela* and *Chloroxylon*).

TYPE—Persian Lilac or Bead-tree (*Melia Azedarach*).

A middle-sized (introduced) tree, with twice-pinnate leaves, and large, much-branched panicles of purplish-white flowers.

Organ.	No.	Cohesion.	Adhesion.
Calyx. *sepals.*	5(-6)	Gamosepalous.	Inferior.
Corolla. *petals.*	5(-6)	Polypetalous.	Hypogynous.
Stamens.	10(-12)	Monadelphous.	Hypogynous.
Pistil. *carpels.*	5	Syncarpous.	Superior.

The Nim, Neem, or Margosa-tree (*Azadirachta indica*) differs from the type in having a three-celled ovary and a one-seeded drupe.

Several valuable timber-trees belong to this Family; amongst the rest, the Indian Satinwood (*Chloroxylon Swietenia*), and the American and West Indian Mahogany and Cedar (*Swietenia* and *Cedrela*).

FIG. 119. *Melia Azedarach*, with a detached drupe (much reduced)

Chloroxylon differs from our type in its free stamens, and (as does also the Mahogany-tree) in a dehiscent, woody, capsular fruit. The fruit of *Azadirachta* is a one-seeded drupe, from the pulp of which an oil is prepared in India. Its bark and leaves are in repute as medicines.

29. Natural Order, *Olacaceæ.*—The Olax Family.

Shrubs or trees, usually with alternate entire leaves, and axillary fascicles spikes or racemes of small flowers.   Ovary one-celled, or imperfectly three- or more-celled.

TYPE—*Olax scandens.*

A climbing spinose shrub, with alternate simple leaves, and short axillary fascicles of small whitish flowers.

Organ.	*No.*	*Cohesion.*	*Adhesion.*
Calyx. *sepals.*	5(-6)	Gamosepalous.	Inferior.
Corolla. *petals.*	5(-6)	Gamopetalous.	Hypogynous.
Stamens.	8(-12)	Octandrous.	Epipetalous.
Pistil. *carpels.*	3	Syncarpous.	Superior.

OBSERVE the "calyx" enlarging (*accrescent*) after flowering-time is over.   Although described as a true calyx consisting of sepals, there is reason to believe that, in this case, it is a cupuliform production analogous to the disk found, for example, around the base of the ovary in the Orange.   Note, also, the abortion of most of the stamens, usually but three bearing perfect anthers, the rest being deeply bifid.

The Family includes several very anomalous Indian climbing plants, differing from the type in their dioecious flowers and other characters, of interest chiefly on botanical grounds.

30. Natural Order, *Ampelideæ.*—The Vine Family.

Usually shrubs climbing by tendrils, with jointed stems,

alternate, simple or tri- to quinque-foliolate leaves, and minute, cymose, greenish flowers.

TYPE—The Grape Vine or any species of *Vitis.*

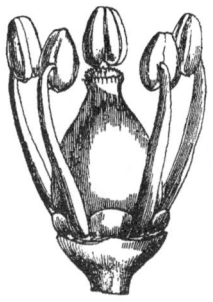
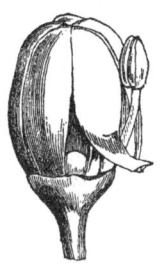

FIG. 120. Flower of Grape Vine after the fall of the petals. The calyx is very short, cup-shaped, and obscurely five-toothed (enlarged).

FIG. 121. Same ; the petals, cohering at their apices, are about to fall.

Usually climbing shrubs, with tendrils, alternate, simple lobed or entire leaves, and panicles (opposite to the leaves) of small, greenish flowers.

Organ.	*No.*	*Cohesion.*	*Adhesion.*
Calyx. *sepals.*	5(4)	Gamosepalous.	Inferior.
Corolla. *petals*	5(4)	Polypetalous.	Hypogynous.
Stamens.	5(4)	Pent-(tetr-)androus.	Hypogynous.
Pistil. *carpels.*	2	Syncarpous.	Superior.

The species of *Leea* represent a Sub-type, differing in the absence of tendrils, their once-, twice-, or thrice-pinnate leaves, and coherent stamens adnate to the petals.

OBSERVE the peduncles and tendrils given off opposite to

the leaves and not as axillary branches. A comparison of the
tendrils and flower-bearing panicles will show that they are
both modifications of the same organ. Each tendril and
panicle is regarded as an axis, the lower portion of which
forms the internode (or internodes), immediately below, of the
main stem. The succeeding internode of the stem is a new
and distinct axis, originating in the axil of the leaf opposite
to which the tendril or peduncle is given off. Observe also
the tendency of the petals to cohere at their apices, so that
they are thrown off as a cap by the expanding stamens.

To numerous varieties of one species of this Family (*V.
vinifera*) we owe nearly all our wine, raisin, and dessert
grapes.

In a curious Malayan genus (*Pterisanthes*) the flowers are
arranged upon a broad, flattened, membranous expansion of
the peduncle, which looks like a monstrous condition.

31. Natural Order, *Sapindaceæ.*—The Soapwort Family.

Usually trees with alternate, pinnate leaves, and inconspicuous, often panicled, polygamous flowers. Stamens free;
ovary three-, four-, or two-celled.

TYPE—The Litchi (*Nephelium Litchi*).

Organ.	No.	Cohesion.	Adhesion.
Calyx. sepals.	4-6	Gamosepalous.	Inferior.
Corolla. petals.	o	o	o
♂ Stamens.	6-10	Usually octandrous.	Hypogynous.
♀ Pistil. carpels.	2(-3)	Syncarpous.	Superior.
Seeds solitary, exalbuminous, arillate.			

A garden tree with shining, pinnate (tri- to septem-folio-late) leaves, and terminal and axillary panicles of numerous small, apetalous, polygamous flowers.

This large Natural Order includes several Indian genera differing much in habit as well as in some of the more technical characters of the flower.   Note especially :

Cardiospermum Halicacabum ; a much-branched climbing herb, with twice-ternate, alternate leaves, umbellate panicles of small irregular flowers, and inflated membranous capsules :

Dodonæa viscosa ; a shrub with narrow, entire, usually undivided, rather viscid leaves, small greenish-yellow, uni-sexual flowers, and winged capsules : and

Maple (*Acer,* any species).   Trees (of the mountains of Northern India) with simple, opposite, usually palmi-veined and lobed leaves, and regular, racemose or corymbose, greenish flowers.

OBSERVE the pulpy aril surrounding the seed of the Litchi, rendering it an esteemed dessert fruit in India and China : the slender tendrils at the end of the peduncle of *Cardiospermum,* and the heart-shaped hilum of its round seeds, to which the name *Cardiospermum* (Heart-seed) refers.

It is generally characteristic of the Family to have the flowers *unsymmetrical,* owing to the number of stamens not corresponding (either the same or as a multiple) with that of the petals and sepals.

The Longan is the fruit of a near ally of the Litchi,— *Euphoria Longana.*   The structure of the fruit is similar to that of the Litchi.   In each of them the lobes of the ovary (carpels), as they mature, become almost or wholly free from each other.   The carpels are one-seeded in both.

Owing to the presence of a saponaceous principle in the fruit of several species of *Sapindus,* the drupes may be used

in lieu of soap.   From this property the name of the Family
is derived.

32. Natural Order, *Anacardiaceæ.*—The Mango Family.

Trees with alternate or opposite, simple or compound
leaves.   Flowers small.   Ovary unilocular (in *Spondias* and
allies, four- or five-celled), with solitary ovules.

TYPE—The Mango (*Mangifera indica*).

FIG. 122. Flower of Mango (*Mangifera indica*), enlarged.

A large tree, cultivated everywhere, with alternate, simple,
lanceolate, shining leaves, and terminal erect panicles of
small, yellowish, polygamous flowers.

Organ.	No.	Cohesion.	Adhesion.
Calyx. *sepals.*	5	Polysepalous.	Inferior.
Corolla. *petals.*	5	Polypetalous.	Hypogynous.
Stamens.	1(-5)	Monandrous.	Perigynous.
Pistil. *carpel.*	1	Apocarpous.	Superior.

OBSERVE the usual abortion of all the stamens except one
in the flower of the Mango: the five or six carpels of
Buchanania, of which only one is perfected : the accrescent

petals in two Malayan genera (*Melanorrhœa* and *Swintonia*), and the accrescent sepals in another (*Parishia*). Note, also, the gum-resin exuded by the bark. Several Indian species of the Family belonging to the genera *Semecarpus* and *Melanorrhœa* yield, either from their bark or fruit, a resinous product, which is often acrid and poisonous. The Indian Marking-nut (*Semecarpus Anacardium*) affords a black corrosive juice used in marking cloth, &c. Mastic is the resin of a shrubby *Pistacia* (*P. Lentiscus*) growing in the Greek Archipelago.

Besides the Mango, which is, *par excellence*, the fruit of India, the Family includes a few other species affording useful and edible fruits. Of the Cashew-nut—the fruit of an American tree—the edible portion is the swollen pear-shaped peduncle of the flower. The kernel of the seed may be eaten when cooked.

Pistachio-nuts are the produce of *Pistacia vera*. *Odina Wodier*, a common tree of the Peninsula, easily propagated by cuttings, is a member of this Family.

33. Natural Order, *Connaraceæ.*—The Connarus Family.

Trees or shrubs, with alternate, compound leaves. Flowers small, regular, in racemes or panicles. Ovary apocarpous.

TYPE—*Connarus pentandrus.*

Organ.	No.	Cohesion.	Adhesion.
Calyx. *sepals.*	5	Gamosepalous.	Inferior.
Corolla. *petals.*	5	Polypetalous.	Hypogynous.
Stamens.	10	Monadelphous.	Hypogynous.
Pistil. *carpel.*	1	Apocarpous.	Superior.

A tree with alternate, pinnate (tri- to septem-foliolate), shining leaves, terminal panicles of small whitish flowers, and short, oblique, one-seeded legumes.

A small Family, represented in India by several genera, but the species are mostly Malayan : a few occur in Bengal and the Peninsula. Their principal interest is in their connecting the large Family Leguminosæ with some of those which precede. They differ principally from Leguminosæ in usually having two or more distinct carpels, in the form and position of the ovules, and in the absence of stipules. The species which we employ as Type, and which is perhaps the most widely-distributed member of the Family in India, is exceptional amongst Connaraceæ in having usually but a single carpel. If there be more carpels in the flower, they are all suppressed excepting one, which forms a short stipitate pod when ripe, very similar to that of some Leguminosæ.

None of the species are of much economic value.

Division—CALYCIFLORÆ.

34. Natural Order, *Leguminosæ.*

Trees, shrubs, or herbs, usually with alternate, compound (sometimes uni-foliolate) leaves. Flowers irregular (except Mimoseæ). Carpel solitary.

This Family of flowering plants is numerically the largest, next to the Composite Family, and includes very many species of great importance to mankind.

Three principal Types of floral structure require to be noted, two of which are, at first sight, very dissimilar ; but in the fruit, seed, and general characters of habit and leafage, the numerous genera of the Family are bound together by a community of sufficiently well-marked characters, so 'that the Order is easily recognised, and may be regarded as a truly natural one.

Peaflower Tribe—*Papilionaceæ.*

TYPE I—The Dhak (*Butea frondosa*).

FIG. 123. Dhak (*Butea frondosa*).

A middle-sized tree, with large, alternate, stipulate, trifoliolate, deciduous leaves, and numerous racemes of handsome, irregular, orange-red, silky flowers.

O 2

Organ.	No.	Cohesion.	Adhesion.
Calyx. *sepals.*	5	Gamosepalous.	Inferior.
Corolla. *petals.*	5	Polypetalous.	Perigynous.
Stamens.	10	Diadelphous.	Perigynous.
Pistil. *carpel.*	1	Apocarpous.	Superior.
Seed solitary, exalbuminous.			

If the Dhak be not at hand, the flower of any Garden Pea (*Pisum*), Bean (*Vicia*), Indigo (*Indigofera*), Gram (*Cicer*), or Lentil (*Ervum*), will answer equally well as Type of the Pea-flower Tribe.

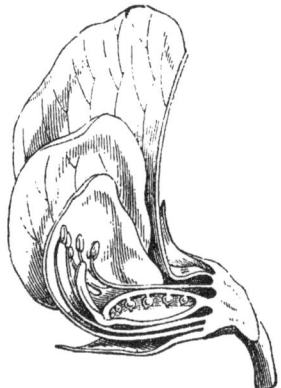

OBSERVE the relation of the petals of the irregular (*papilio-naceous*) corolla of the Peaflower Tribe to each other. There is a large upper petal which embraces the rest in the bud; this is the standard: two lateral petals, called *wings*, and two, usually more or less coherent by their lower margins, forming the *keel*, which enclose the stamens and pistil.

FIG. 124. Section of flower of Garden Pea.

Cassia Tribe—*Cæsalpinieæ.*

TYPE 2—*Cassia Fistula.*

A small spreading tree, with alternate, pinnate, glabrous, stipulate, deciduous leaves, pendulous racemes of bright-yellow, fragrant, irregular flowers, and long cylindrical indehiscent pods. (Corolla, *petals imbricated, upper petal inside in bud.*)

In the absence of *C. Fistula* any common Cassia, as the annual species *C. Sophora,* or *C. Tora,* or the ornamental garden-shrub *Poinciana pulcherrima,* will serve.

Mimosa Tribe—*Mimoseæ.*
TYPE 3—Humble Plant (*Mimosa pudica*).

A prickly, hairly, perennial garden herb, with alternate, digitate-pinnate, sensitive, stipulate leaves, and axillary, pedunculate heads of small, pale-purple, regular, polygamous flowers. (Corolla *regular, petals valvate in bud.*)

OBSERVE the stamens in the type of the Pea-flower Tribe. Nine cohere into a bundle by their filaments : one (the upper one next to the *standard*) remaining free. In the Sunn Hemp (*Crotalaria*) and several other genera the stamens are monadelphous ; in *Abrus* there are only nine stamens ; while in Sophora, as in Cassia and Mimosa, types of the second and third Tribes, they are all free. In Mimosa and its allies they are often indefinite.

FIG. 125. Diadelphous stamens of Pea.

Compare the fruit (*legumes*) of any Pea, Bean, or Haricot (dehiscent, two-valved) ; Dhak (indehiscent, one-seeded) ; Crotalaria (inflated legume) ; *Æschynomene, Desmodium, Alysicarpus, Uraria* (articulated, separating into distinct one-seeded articles) ; *Dalbergia* (thin, flat, indehiscent) ; *Pterocarpus,* Sanders-wood (winged, indehiscent) ; *Sophora* (cylindrical, often narrowed here and there but not jointed, indehiscent) ; *Arachis,* the Ground Nut (indehiscent, one- to three-seeded,

matured underground); *Cassia Fistula* (cylindrical, indehiscent, divided by spurious transverse plates developed
from the endocarp into one-seeded cells); *Guilandina* (dehiscent, prickly); *Entada* (huge, sword-shaped, indehiscent,
several feet in length, the valves separating into one-seeded
articles, which fall from the thickened and consolidated
sutures); *Tamarindus,* the Tamarind (indehiscent, with the
seeds immersed in an acid pulp); *Albizzia Lebbek* (thin and
papery legumes, breaking up into one-seeded articles)

Fig. 126. Tamarind (*Tamarindus indica*): a detached legume to the left. About
one-third natural size.

The seeds of Leguminosæ are generally exalbuminous,
but in a few genera, as *Cassia,* some albumen is present.

Of species serviceable to man we can here notice but few.

Of Timber-trees, the more important are the cabinet Rosewoods, afforded by Brazilian species of *Dalbergia.* The same genus includes some valuable Indian timber-trees, especially *D. Sissoo.* The wood of the Tamarind is sometimes used in cabinet-work.

Of Food-producing plants, the more important are the Pea, Bean, Lentils, Haricots, and Kidney-beans, Ground Nut (*Arachis*), Gram (*Cicer*), and Pigeon Pea (*Cajanus*).

Of Dyes we have, of first importance, the Indian product Indigo, obtained by decomposing the herbage of species of *Indigofera;* Red Sanders-wood, afforded by *Pterocarpus santalinus,* and Sappan-wood by *Cæsalpinia Sappan,* both large trees of the Peninsula; Logwood, by the Central American *Hæmatoxylon.*

Crotalaria juncea, the Sunn Hemp, is an annual, largely grown in India for the sake of the tenacious hemp-like fibre afforded by the bark of its long shoots.

Many of the Leguminosæ are serviceable in medicine, and some afford resins, balsams, or astringent gums, as the Dhak, which is one of the Indian lac-producing trees, and the Catechu (*Acacia Catechu*); the latter a powerful astringent, the resin of which is obtained by boiling the wood. It is exported to England for the use of tanners.

The familiar irritability of the compound leaves of the Humble and Sensitive plants is but an extreme case of the condition (called the sleep) exhibited by many of the Leguminosæ, the Sorrels, &c., the leaflets of which fold together in the evening and remain closed until morning. The Indian species of the genus *Acacia,* closely allied to Mimosa, are characterised by highly compound (twice pinnate) leaves. It is remarkable that in Australia, where this genus has its head-quarters, a large proportion of the species have leaves wholly destitute of a blade, being reduced to a petiole, which, in order to compensate for the deficiency, is much flattened and leaf-like, serving the purpose, as to function, of an ordinary leaf-blade.   Such flattened petioles are called phyllodes.   Their true nature is plainly ascertained by finding the compound blade sometimes developed upon the flattened phyllode, as is the case often in seedling Acacias.

Some of these phyllodineous Acacias have been introduced into Southern India, where they are said to hold their own, and appear likely to become naturalized.

35. Natural Order, *Rosaceæ.*—The Rose Family.

Trees, shrubs, or herbs, with alternate, entire or divided leaves.   Flowers regular.   Ovary free, or adherent to the calyx-tube (when the pistil becomes apparently syncarpous).

TYPE—Any species of Bramble (*Rubus*).

Scrambling, prickly shrubs, with alternate, digitate or trifoliolate leaves, and terminal panicles of regular white or rose-coloured flowers.

Organ.	*No.*	*Cohesion.*	*Adhesion.*
Calyx. *sepals.*	5	Gamosepalous.	Inferior.
Corolla. *petals.*	5	Polypetalous.	Perigynous.
Stamens.	∞	Polyandrous.	Perigynous.
Pistil. *carpels.*	∞	Apocarpous.	Superior.
Seeds solitary, exalbuminous.			

This Family, of great importance in orchard and garden culture of the temperate zone, is represented by several genera in India, which do not however include many species of much economic importance. The variety, chiefly in the arrangement and number of the carpels, and their position relative to the tube of the calyx, may be reduced to three Sub-types, each represented in India either by native or cultivated species.

Bramble (*Rubus*) may serve as type of the Sub-order Roseæ, marked by numerous free carpels.

FIG. 127. Vertical section of flower of Bramble.

Cherry, or any species of *Prunus* or *Pygeum*, of the Sub-order Drupaceæ, marked by a single free carpel, drupaceous in fruit.   And the

Loquat (*Eriobotrya*), *Photinia*, or Apple (*Pyrus*), of the Sub-order Pomaceæ, marked by one or more carpels adherent to the calyx-tube, so that the ovary is inferior.

Most of the species of this large Family agree in their perigynous stamens, which are usually indefinite ; polypetalous corolla, prone to become "double" at the expense of the stamens, as in Rose and *Kerria;* and the essentially apocarpous pistil.

In Pomaceæ, if the fruit be cut across, it will be seen that the carpels do not cohere *inter se*, though pressed and bound together by the succulent enlargement of the so-called calyx-tube.

OBSERVE the *prickles* of Bramble and Rose, differing from *spines* (page 72) in being processes of the bark and not developments of the axis. Compare the arrangement of the

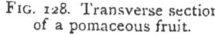

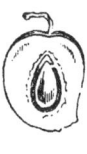

FIG. 128. Transverse section of a pomaceous fruit.   FIG. 129. Fruit of Bramble (*Rubus*).   FIG. 130. Single drupel of Bramble.

carpels in the genera Bramble and Rose. In both the carpels are wholly free from each other, as well as from the calyx-tube ; but in the latter they are arranged upon the inside of

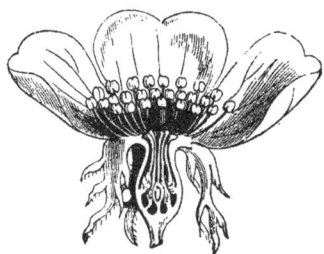

FIG. 131. Fruit of Strawberry.   FIG. 132. Vertical section of flower of Rose (*Rosa*).

the urn- or flask-shaped tube of the calyx (regarded by some botanists as a hollowed receptacle), while in the former they are disposed upon a conical projecting receptacle. In the European Strawberry the arrangement of the carpels is similar to that of the Bramble, but the receptacle becomes inordinately enlarged and succulent, bearing the small dry carpels (which are commonly and incorrectly called seeds) upon its surface.

In the species of *Spiræa*, some of which are Himalayan, the fruit-carpels are several-seeded follicles.

To this Family belong the following invaluable fruits, most of them grown in perfection only in temperate climates :—Apple, Pear (*Pyrus*), Quince (*Cydonia*), Medlar (*Mespilus*), Almond, Peach, Nectarine (*Amygdalus*), Apricot (*Armeniaca*), Cherry and Plum (*Prunus*), Strawberry (*Fragaria*), Raspberry (*Rubus*), and Loquat (*Eriobotrya*),—the last a Japanese tree, grown in Indian gardens.

Excellent Otto of Rose is obtained in India from the petals of sweet-scented Roses. Most of that sold in London is adulterated with the oil of an Indian grass (*Andropogon*).

36. Natural Order, *Combretaceæ.*—The Combretum Family.

Trees or shrubs, with opposite or alternate simple leaves. Flowers with or without petals. Ovary wholly inferior, one-celled, with pendulous ovules.

Type—The Rangoon Creeper (*Quisqualis indica*).

Organ.	No.	Cohesion.	Adhesion.
Calyx. *sepals.*	5	Gamosepalous.	Superior.
Corolla. *petals.*	5	Polypetalous.	Perigynous.
Stamens.	10	Decandrous.	Perigynous.
Pistil. *carpel.*	1 (?)	Apocarpous.	Inferior.

A more or less climbing pubescent shrub, with opposite or nearly opposite, simple, entire leaves, and terminal and axillary lax spikes of slender, reddish flowers.

With *Quisqualis* compare *Terminalia Catappa*—a large ornamental tree, common in gardens, with simple entire leaves clustered towards the ends of the branches, and small, whitish, apetalous, polygamous flowers, in axillary, simple, erect racemes.

OBSERVE the one-celled ovary with pendulous ovules, characteristic of the Family : the spirally-twisted cotyledons of the single seed of the Terminalias.

The kernels of the fruit of *T. Catappa* are in much esteem. They are eaten as Almonds. The fruits—Myrobolans—of other species of *Terminalia* (*T. Bellerica* and *T. Chebula*) are astringent, and are exported to Britain for the use of tanners, &c. Some of the Terminalias are valuable timber-trees.

37. Natural Order, *Celastraceæ.*—The Spindle-tree Family.

Shrubs or trees, with opposite or alternate simple leaves and minute flowers. Ovary more or less immersed in a disk. Stamens alternate with the petals and equal in number, or only three.

TYPE—*Celastrus paniculatus.*

Organ.	No.	Cohesion.	Adhesion.
Calyx. *sepals.*	5	Gamosepalous.	Inferior.
Corolla. *petals.*	5	Polypetalous.	Perigynous.
Stamens.	5	Pentandrous.	Perigynous.
Pistil. *carpels.*	3	Syncarpous.	Superior.

A climbing shrub, with alternate, simple, serrate leaves, and terminal panicles of numerous small yellow flowers.

The Indian species of *Hippocratea* and *Salacia* represent a subordinate type, differing in having triandrous flowers and exalbuminous seeds.

OBSERVE the yellow pulpy covering (*arillus*) of the seed, exposed, when the capsules split loculicidally, while still hanging on the plant. The arillus is a cellular investment growing more or less over the seed as it matures, either from the funicle (the pedicel by which the ovule is attached to the placenta) or from the micropyle.

38. Natural Order, *Rhamnaceæ.*—The Buckthorn Family.

Trees or shrubs, with alternate simple leaves and minute flowers. Stamens opposite to the petals and equal in number.

TYPE—The Jujube (*Zizyphus Jujuba*).

A thorny shrub or small tree, with alternate, three-nerved, sub-distichous leaves, shining above, white or rusty-downy beneath; axillary, umbellate fascicles of small, greenish flowers; and yellow, drupaceous, edible fruits.

Organ.	No.	Cohesion.	Adhesion.
Calyx. *sepals.*	5	Gamosepalous.	Superior.
Corolla. *petals.*	5	Polypetalous.	Perigynous.
Stamens.	5	Pentandrous.	Perigynous.
Pistil. *carpels.*	2	Syncarpous.	Inferior.

Although the ovary at the time of flowering is inferior, or half-inferior, being immersed in the fleshy disk, the fruit is wholly free and superior.

OBSERVE the stamens opposite to the minute petals, a character by which this Family may be distinguished from the preceding.

The fruit of the Type is well known as the jujube. That of *Z. Lotus* is said to have been eaten by an ancient people of North Africa, who were hence called, according to some authorities, Lotophagi. The berries of some species of Buckthorn (*Rhamnus*) afford a yellow dye, and, treated with alum, the pigment called "sap-green."

39. Natural Order, *Melastomaceæ.*—The Melastoma Family.

Herbs or shrubs, with opposite, entire, three- (or more-) nerved leaves. Petals twisted in bud. Stamens ten or fewer, perigynous.

TYPE—*Melastoma malabathricum.*

An erect, shrubby plant, or small tree, with opposite,

FIG. 133. *Melastoma malabathricum* (reduced).

three- to five-nerved, entire leaves rough with adpressed bristles, and regular, large, terminal, red flowers.

Organ.	No.	Cohesion.	Adhesion
Calyx. *sepals.*	5	Gamosepalous.	Half-inferior.
Corolla. *petals.*	5	Polypetalous.	Perigynous.
Stamens.	10	Decandrous.	Perigynous.
Pistil. *carpels.*	5	Syncarpous.	Half-superior.

A thoroughly tropical Family, of very little economic use, though generally characterised by beautiful flowers. The head-quarters of the Family is in Brazil.

OBSERVE the three strong nerves, almost invariably present, of the opposite leaves : the curious structure of the stamens, and the mode in which the anthers are tucked into little pockets between the ovary and calyx-tube while enclosed in the bud. In our Type there are five long and five short

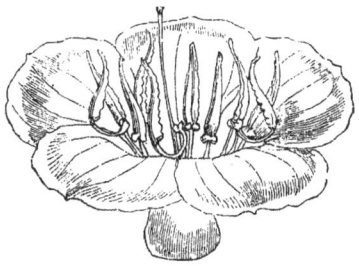

FIG. 134. Flower of *Melastoma malabathricum.*

stamens; the anthers of the long ones are prolonged and curved below the cells, and at the point where they join the filament they give off two little horns.

The Family derives its name from the genus *Melastoma*, the name of which signifies "black-mouth," from the ripe berries, which are edible, dyeing the mouth black. The fruit of our Type species affords a purple dye.

40. Natural Order, *Myrtaceæ.*—The Myrtle Family.

Trees or shrubs, usually with opposite, entire leaves marked with translucent dots. Stamens indefinite. Ovary adherent, with axile placentas.

TYPE.—The Jambolan (*Syzygium Jambolanum*).

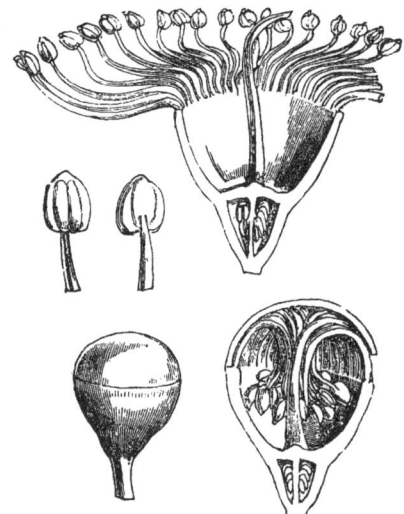

FIG. 135. Jambolan (*Syzygium Jambolanum*): showing a bud; the same in vertical section; an expanded flower after the fall of the calyptrate corolla; and two detached stamens.

A tree with opposite, entire, shining, exstipulate leaves, and trichotomous panicles of small white flowers from the axils of fallen leaves.

Organ.	No.	Cohesion.	Adhesion.
Calyx. *sepals.*	4	Gamosepalous.	Superior.
Corolla. *petals.*	4	Polypetalous.	Perigynous.
Stamens.	∞	Polyandrous.	Perigynous.
Pistil. *carpels.*	2	Syncarpous.	Inferior.

OBSERVE the leaves of any species of the Family; they are marked with translucent, glandular dots, like those of the Oranges, and an intra-marginal vein (*i.e.* a vein running parallel with and just within the margin): the deciduous petals of the Jambolan and the Clove (*Caryophyllus aromaticus*); they are thrown off on expansion of the flower as the stamens unfold. In some other genera the lobes of the calyx-limb are coherent, and are thrown off in a similar way as a lid or *calyptra.* This is well shown in the large Australian genus *Eucalyptus.*

The Myrtle Family is chiefly tropical, abounding in Brazil, India, and Australia. A single species, the Common Myrtle (*Myrtus communis*) is native in Europe.

Eucalyptus, an Australian genus, includes many gigantic timber-trees, known as Stringy-barks, Iron-barks, and Gum-trees. Some of the species have been introduced into the South of India, and promise to succeed.

Cloves are the dried, unopened, flower-buds of *Caryophyllus*, referred to above. They are grown principally in the Indian Archipelago, Africa, and the West Indies. Clove pepper, another spice of the Family, is the dried unripe fruit of *Pimenta vulgaris*, a tree extensively planted in Jamaica.

P

The Guava (*Psidium Guava*) is much cultivated through-out the tropics for the sake of its delicious fruit.

The Pomegranate (*Punica granatum*), cultivated in India, is an exceptional member of this Family. The structure of its fruit is puzzling, owing to the development of two distinct whorls of carpels, which become completely con-solidated. It is believed to have been originally a native of Western Asia, and not of Carthage, as its name would denote (*Malum granatum*).

Close allies of the Myrtles are the so-called Monkey-pots (*Lecythideæ*) of tropical America.

They are remarkable for the transverse dehiscence of their large, woody capsules. Brazil-nuts are the seeds of a species of this Tribe (*Bertholletia excelsa*), as are, also, Sapucaja-nuts (*Lecythis*, sp.).

None of the Myrtle Family are blue-flowered.

41. Natural Order, *Rhizophoreæ*.—The Mangrove Family.

Trees or shrubs with opposite entire coriaceous leaves. Calyx-teeth valvate. Petals often fringed. Ovary more or less adherent.

TYPE—*Rhizophora mucronata* (or any other species of Mangrove).

Organ.	No.	Cohesion.	Adhesion.
Calyx. *sepals.*	4	Gamosepalous.	Superior.
Corolla. *petals.*	4	Polypetalous.	Perigynous.
Stamens.	8	Octandrous.	Perigynous.
Pistil. *carpels.*	4	Syncarpous.	Inferior.

Littoral, swamp trees, often with branching, adventitious roots, opposite, entire, thick, smooth leaves, axillary forked peduncles bearing small, whitish flowers, and the seed germinating before falling from the parent.

OBSERVE the tendency, in species growing between tide-marks, to develope numerous adventitious roots, which serve to secure a firm hold of the ground : the fringed petals of several genera : the growth of the radicle of the embryo of the single seed while still contained in the fruit and before it falls. It grows to a length, sometimes, of several feet, reaching the mud, and throwing out branch-rootlets from its base before parting from the parent tree.

Compare with the Type-species any *Bruguiera*, with the calyx about twelve-lobed, the same number of petals, and double the number of stamens.

The bark of the Mangrove is astringent, and may be used in tanning.

42. Natural Order, *Onagraceæ.*—The Evening-Primrose Family.

Herbs with alternate or opposite, simple (sometimes divided if submerged) leaves. Calyx-teeth usually four ; valvate. Ovary adherent, two- to four-celled.

TYPE—*Jussieua repens.*

Organ.	No.	Cohesion.	Adhesion.
Calyx. *sepals.*	5	Gamosepalous.	Superior.
Corolla. *petals.*	5	Polypetalous.	Perigynous.
Stamens.	10	Decandrous.	Perigynous.
Pistil. *carpels.*	5	Syncarpous.	Inferior.

A floating or creeping annual herb, with alternate, entire, obovate leaves, and axillary, solitary, pedunculate, reddish or yellow-white, regular flowers.

This common water-plant is exceptional in having its parts usually in fives ; the flowers of the Family are almost invariably *tetramerous*, *i.e.* with the parts in fours.

OBSERVE the floats of cellular tissue attached to the submerged nodes of the *Jussieua :* the valvate æstivation of the calyx-lobes and twisted æstivation of the petals.

In the garden species of the ornamental South American genus *Fuchsia* the calyx is beautifully coloured, as well as the corolla. In some species of *Fuchsia* the corolla is very small, or wholly absent.

Onagraceæ are chiefly plants of temperate regions, and they are not a conspicuous feature in India.

Nearly allied to this Family is the small aquatic Order Halorageæ. It includes the Water Chestnuts (*Trapa bispinosa* and *T. bicornis*), important food-plants in Thibet, N.W. India, and China. The fruit of *T. bicornis* resembles the head of an ox in miniature.

43. Natural Order, *Lythraceæ.*—The Loosestrife Family.

Trees, shrubs, or herbs, with opposite (or alternate) simple leaves. Ovary free.

TYPE—*Lagerstrœmia indica.*

Organ.	No.	Cohesion.	Adhesion.
Calyx. *sepals.*	6	Gamosepalous.	Inferior.
Corolla. *petals.*	6	Polypetalous.	Perigynous.
Stamens.	18(-30)	Polyandrous.	Perigynous.
Pistil. *carpels.*	6	Syncarpous.	Superior.

A showy garden shrub, with opposite or alternate entire leaves and white or purple flowers in small, terminal, racemose panicles.

OBSERVE the six outer stamens longer than the rest. In a common English species of this Family (*Lythrum Salicaria*) there are three forms of flower, each characterised by the same relative length of the style and stamens. In one, the style is shorter than the six short stamens; in another, it is intermediate in length between the short and long stamens; and in the third it is longer than the long stamens. These differences Mr. Darwin shows to be designed to favour the crossing of the stigma by the pollen of other flowers, especially by flowers which have stamens corresponding in length to the style of the flower to be fertilized. Observations upon Indian species with respect to similar dimorphic (or trimorphic) conditions are much needed. Observe also the slight infolding of the dorsal sutures of the carpels forming the syncarpous ovary of the Type-species.

A second magnificent species of *Lagerstræmia* (*L. reginæ*) is a native of the Indian Peninsula. It grows to a large size and affords useful timber. *Grislea tomentosa*, a common Indian shrub, with racemes of red flowers, used as a dye and in medicine, and with a persistent red calyx enclosing the capsule, is a member of this Family. Henna, used by Egyptian ladies to dye their nails and the palms of their hands a reddish-brown colour, is obtained from the leaves of *Lawsonia inermis*. The nails of mummies are sometimes found dyed with it.

44. Natural Order, *Cucurbitaceæ.*—The Gourd Family.

Climbing or prostrate herbs, with alternate leaves and lateral tendrils. Flowers unisexual. Ovary inferior.

TYPE—Common Gourd, *Cucurbita maxima;* or
Cucumber, *Cucumis sativus;* or
Bottle Gourd, *Lagenaria vulgaris;* or
Water Melon, *Citrullus vulgaris.*

Annual, creeping, usually rough or hairy herbs, with
alternate, palmi-veined and lobed leaves, extra-axillary
tendrils, and solitary, axillary, unisexual, regular, yellow
or white flowers.

Organ.	*No.*	*Cohesion.*	*Adhesion.*
Calyx. *sepals.*	5	Gamosepalous.	Superior.
Corolla. *petals.*	5	Poly- or gamo-petalous.	Perigynous.
♂ Stamens.	3	Triandrous.	Perigynous.
♀ Pistil. *carpels.*	3	Syncarpous.	Inferior.
Seeds ∞, exalbuminous.			

OBSERVE the tendrils, regarded as modified leaves of
axillary or extra-axillary shoots, the internodes
of which are undeveloped, the tendril-leaves
being reduced as it were to their principal
veins, which serve as feelers and hold-fasts:
the petals, in some genera free, in others
coherent, forming a bell-shaped corolla: the
three stamens, the anther of one stamen being
one-celled, of the others, two-celled. The
anther-cells are remarkably sinuous, being
twisted up and down like the letter S. In
Cucurbita the anthers are coherent. This
very important tropical Family, which includes many species cultivated in India, is

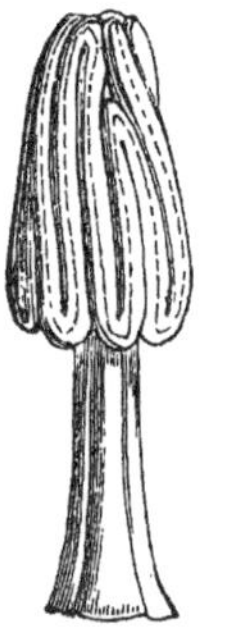

FIG. 136 Mon-adelphous sta-mens with sin-uous anthers.

remarkable for the great variety, in form, of the fruit. It is often very variable in the same species, as in the Gourd and Bottle Gourd. The firm outer pericarp, hollowed out, serves a variety of purposes; it is often ornamented by painting or burnt lines.

Many species are intensely bitter, and some are dangerous poisons, as the Colocynth (*Citrullus colocynthus*). Besides the species named as types, the following are employed in India as food-plants (most of them are cultivated):— *Karivia umbellata, Momordica charantia, M. dioica, Luffa pentandra, L. fœtida, Benincasa cerifera, Trichosanthes anguina* (the Snake Gourd), *Coccinia grandis,* and others.

Allied to the Gourds, and scrambling or climbing by the aid of tendrils, are the Passion-flowers (*Passifloreæ*), an Order principally South American, but represented in India by a few native species, and several cultivated in gardens for the sake of their beautiful purple, scarlet, or greenish-white flowers. They differ from the Gourds in their herma-phrodite flowers, superior, often stalked ovary, and the beautiful corona of filiform appendages arising from the tube of the calyx, as well as in other characters. A few Passion-flowers yield an eatable fruit, as the Granadilla (*P. quadrangularis*) of the West Indies.

OBSERVE the remarkable habit and the structure of the sulphur-yellow flowers of the Prickly Pear (*Opuntia Dillenii*); the representative, very widely naturalized in India, of an almost exclusively American Family, the Cactaceæ. This Family is characterised by remarkable succulence, and by a tough impervious skin or epidermis, which checks undue evaporation of the juices in the arid climates to which the group is principally confined. Very few possess developed leaves, the succulent epidermis of the stem performing the function of these organs.

There are three principal modifications in the form of the stem amongst the leafless Cactaceæ, viz. the Columnar, the Globular, and the Jointed or Lobed. The Indian Prickly Pear belongs to the last type. Note the gradual passage from bracts to petals ; both sepals and petals, as well as stamens, being indefinite.

45. Natural Order, *Begoniaceæ.*—The Begonia Family.

Succulent herbs, with oblique, usually alternate, leaves. Flowers unisexual. Ovary adherent, three-celled.

Type—*Begonia laciniata* (or any other species of Begonia).

Herbs with alternate, stipulate, obliquely cordate, five- to seven-lobed leaves, and axillary, two-flowered peduncles of pale-pink unisexual flowers.

Organ.	*No.*	*Cohesion.*	*Adhesion.*
Perianth. *leaves.*	4(-5)	Gamophyllous.	Superior.
♂ Stamens.	∞	...	...
♀ Pistil. *carpels.*	3	Syncarpous.	Inferior.

Observe the almost invariable obliquity of the leaves, the midrib dividing them unsymmetrically : the sinuous stigmas : the forked placentas, with indefinite ovules.

The species of this small Family nearly all belong to the genus Begonia. They are very widely spread through the tropics, excepting in Africa, where but comparatively few species have been met with. Few species are applied to any use, excepting for ornamental purposes, for which several Indian species are well fitted by their beautifully variegated leaves. The variegation of Begonias, as in *Cissus discolor*, arises from the presence of a film of air under the epidermis wherever the surface appears silvery. The epidermis is an interesting microscopic object, owing to the stomates being frequently collected in clusters.

46. Natural Order, *Crassulaceæ.*—The Stonecrop Family.

Herbs or shrubs, usually with fleshy leaves. Pistil nearly apocarpous. Ovary superior.

Type—*Bryophyllum calycinum.*

An erect, shrubby (introduced) plant, with opposite, simple or compound, succulent leaves, and large terminal panicles of pendulous, greenish-purple, regular flowers.

Organ.	*No.*	*Cohesion.*	*Adhesion.*
Calyx. *sepals.*	4	Gamosepalous.	Inferior.
Corolla. *petals.*	4	Gamopetalous.	Perigynous.
Stamens.	8	Octandrous.	Epipetalous.
Pistil. *carpels.*	4	Apocarpous.	Superior.

OBSERVE the inflated calyx. In some genera of the Family there are as many as twenty sepals. Crassulaceæ are abundant at the Cape of Good Hope, and many species grow in dry situations in temperate countries. They are generally characterised by fleshy leaves.

In Britain, as well as in the Himalaya, many species of the typical genus Stonecrop (*Sedum*) occur, differing from the type which we have employed in having a polypetalous corolla, as well as in habit.

The leaves of *Bryophyllum*, when placed on moist soil, produce young plants from the notches on their margin. This is supposed to illustrate the development of ovules, which are normally "buds" borne upon the margin of carpellary leaves.

47. Natural Order, *Saxifrageæ.*—The Saxifrage Family.

Herbs, shrubs, or trees with alternate or opposite simple leaves. Ovary more or less adherent. Stamens usually ten or fewer.

TYPE—Garden Hydrangea (*Hydrangea hortensis*).

A cultivated shrubby (Chinese) plant, with opposite, simple, serrate leaves, and large, dense, cymose clusters of pale, rose-coloured flowers (principally neuter).

Organ.	*No.*	*Cohesion.*	*Adhesion.*
Calyx. *sepals.*	4(-5)	Gamosepalous.	Superior.
Corolla. *petals.*	4(-5)	Polypetalous.	Epigynous.
Stamens.	4(-5)	Pentandrous.	Epigynous.
Pistil. *carpels.*	2(-4)	Syncarpous.	Inferior.
Seeds indefinite, albuminous.			

This temperate Family is referred to here partly on account of a few tropical Indian representatives, and partly to direct attention to the structure of the flowers of the species which we employ as type. In the Garden Hydrangea the lobes of the calyx are prone to enlarge inordinately at the expense of the essential organs of the flower, so that they nearly all become barren, or neuter. The pale-pink colour of this flower is said to pass into a blue when the soil in which it is raised contains oxide of iron.

Several species of the typical genus *Saxifraga* grow at great elevations in the Himalaya; others are at home in Arctic regions.

Note the Sundews (*Drosera*), regarded as a Tribe of this Family: low, glandular-viscid herbs, usually growing in boggy situations. Some species exhibit a low sensibility or irritability in the leaves, which curl upon particles placed on their glandular hairs. Is this the case with Indian species? And do they appear to discriminate between organic and inorganic matter offered to them?

The Venus' Fly-trap (*Dionæa*) of the South United States affords one of the most remarkable instances of irritability in the vegetable kingdom. The leaves close instantaneously when a fly touches one of three irritable hairs placed upon each lobe of the leaf.

48. Natural Order, *Umbelliferæ.*—The Umbellate Family.

Herbs with hollow stems and sheathing, often dissected, leaves, and small umbellate flowers. Petals and stamens five, epigynous. Carpels two, dry when ripe.

TYPE—either cultivated Fennel (*Fœniculum*); or
Dill (*Anethum*); or
Carrot (*Daucus*); or
Coriander (*Coriandrum*).

Herbaceous plants, with erect, hollow stems, alternate, sheathing, much divided leaves, and small flowers in terminal compound umbels.

FIG. 137. Vertical section of flower of Umbelliferous plant, showing the inferior
two-celled ovary with one ovule in each cell.

Organ.	No.	Cohesion.	Adhesion.
Calyx. *sepals.*	5	Gamosepalous	Superior.
Corolla. *petals.*	5	Polypetalous.	Epigynous.
Stamens.	5	Pentandrous.	Epigynous.
Pistil. *carpels.*	2	Syncarpous.	Inferior.
Seeds solitary, albuminous.			

A very large Family in Europe and temperate Asia, but
of small importance in Indian botany.

OBSERVE the limb of the calyx usually reduced to a mere
rim (then said to be *obsolete*) : the tendency, in the outer
flowers of the umbels, to become irregular, the outside petals
being larger, recalling the relation of the ray and disk
florets in Compositæ : the inflexed tips of the petals : the
ripe fruit, separating into its two carpels, which often remain
suspended from a slender stalk (*carpophore*) in the middle.

The face of union of the two carpels is called the *commissure*. In the substance of the thin pericarp are often found minute, longitudinal canals, containing essential oil. These are the *vittæ*. They may be found, when present, by making a careful transverse section of the ripe fruit.

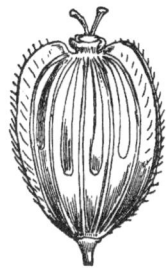 

FIG. 138. One of the fruit-carpels of *Heracleum*, showing four vittæ on the dorsal face.

FIG. 139. Transverse section of same, showing the position of the vittæ ; also the embryo surrounded by copious albumen.

In *Hydrocotyle* the umbels are simple. *H. asiatica* is common in swamps in India and tropical Africa, occurring also in South America and Australia. This wide distribution is shared by numerous plants which affect similar situations, growing either wholly or partially in water probably on account of the facilities afforded to their dispersion by water-fowl.

The carpels of Umbelliferæ are usually marked by lines, ridges, or wings answering to the dorsal and sutural lines of the cohering (and adhering) sepals. In Carrot the ridges are bristly, in *Prangos* winged. The latter is common in Thibet, where it serves as a valuable food for sheep.

The Family includes many dangerous plants, as well as some valuable esculents. Some species afford aromatic fruits or medicinal resins. Hemlock (*Conium*), the State poison of the Athenians, is one of the foremost amongst the dangerous species.   It is useful in medicine.   Cultivated in temperate countries, besides the species named as types, are : Parsnips (*Pastinaca*), Celery (*Apium*), Parsley (*Petroselinum*), Caraways (*Carum*), and Anise (*Anisum*).

49. Natural Order, *Araliaceæ.*—The Ivy Family.

Shrubs, trees, or herbs, with alternate, simple or compound leaves.   Petals and stamens epigynous.   Carpels two or more.

TYPE—*Paratropia venulosa.*

A small tree, with alternate, petiolate, digitate, smooth, veiny leaves, and numerous umbels of very small flowers, arranged in terminal panicles.

Organ.	No.	Cohesion.	Adhesion.
Calyx. *sepals.*	5	Gamosepalous.	Superior.
Corolla. *petals.*	5	Polypetalous.	Epigynous.
Stamens.	5	Pentandrous.	Epigynous.
Pistil. *carpels.*	5	Syncarpous.	Inferior.

This Family may be regarded as representing the Umbellifers in the tropics.   They agree in many technical characters, but generally differ either in the increased number of carpels cohering to form the syncarpous ovary, or in their woody habit.   There are many Natural Orders which include much greater variety of structure than Araliaceæ and Umbelliferæ together, and that in respect of the same organs, but the ordinal characters of the Umbellifers derive their systematic importance from their constant prevalence through an immense number (about 1,500) of species, so that botanists find it convenient to treat Araliaceæ as a distinct Family, although their technical differences are of small absolute importance.

The Ivy is the sole representative of the Family in Britain. The same very variable species is native in the Himalaya and Khasia mountains. The Ginseng root, highly prized by the Chinese as a restorative medicine, and sometimes sold at from 20 to 250 times its weight in silver, is the produce of an herbaceous species of *Panax*, native in temperate Eastern Asia. Rice paper is the pith of a shrubby *Aralia* growing in the island of Formosa. It is cut into small sheets from the thick cylinder of pith, with long knives.

50. Natural Order, *Loranthaceæ.*—The Mistletoe Family.

Parasitical shrubs, with opposite or alternate leaves. Stamens opposite perianth-lobes. Ovary inferoir. Fruit one-seeded.

TYPE—*Loranthus bicolor.*

A woody parasite, with opposite entire leaves and axillary racemes of slender, showy, scarlet and green flowers.

Organ.	*No.*	*Cohesion.*	*Adhesion.*
Calyx. *sepals.*	0	Represented by an adherent cup-shaped disk.	
Corolla. *petals.*	5	Gamopetalous.	Epigynous.
Stamens.	5	Pentandrous.	Epipetalous.
Pistil. *carpels.*	3	Syncarpous.	Inferior.
Seeds solitary, albuminous.			

OBSERVE the intimate attachment of the parasite to its prey, best seen by making a section through the plane of union. The wood of the parasite and that of the stock, although in most intimate union, do not intermix or shade off into each other. Observe, also, the corolla, deeply split on one side, in some species quite regular : the inferior ovary, buried in the disk which grows up around and adherent to it; the single erect ovule which it contains is entirely adnate with the walls of the one-celled ovary, so that if the latter be cut across at the time of flowering, it appears to be quite solid.

Several species of Mistletoe (*Viscum*), both with and without leaves, are common parasites in India. They differ from Loranthus in their minute petals, and in the adnate anthers, dehiscing by numerous microscopic pores.

Division—COROLLIFLORÆ.

* *Ovary inferior*

51. Natural Order, *Rubiaceæ.*—The Peruvian-Bark Family.

Trees, shrubs, or herbs, with opposite simple leaves and interpetiolar stipules. (In some herbs the leaves are verticillate.)

TYPE—*Ixora coccinea.*

Organ.	No.	Cohesion.	Adhesion.
Calyx. *sepals.*	4	Gamosepalous.	Superior.
Corolla. *petals.*	4	Gamopetalous.	Epigynous.
Stamens.	4	Tetrandrous.	Epipetalous.
Pistil. *carpels.*	2	Syncarpous.	Inferior.
Seeds solitary in each cell, albuminous.			

A commonly cultivated (Chinese) garden shrub, with opposite, entire leaves, interpetiolar stipules, and terminal dense clusters of bright scarlet flowers.

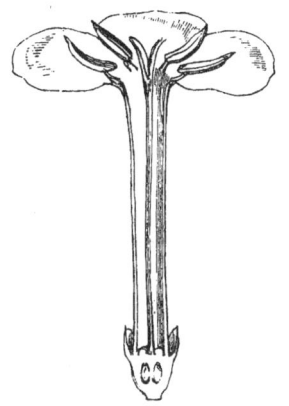

FIG. 140. Vertical section of flower of *Ixora.*

As Sub-types, note—

Gardenia florida of gardens, with one-celled ovary and numerous seeds ; and

Rubia cordifolia, Munjeet or Indian Madder, with four leaves in a whorl (two smaller), a two-celled ovary, and one seed in each cell.

The *Gardenia* is often "double," from a multiplication of the cohering petals and change of stamens into petals. This "doubling" of flowers is much more frequent among polypetalous than amongst gamopetalous Families. It is generally due both to transformation of the stamens into petals and to actual increase in number of parts.

The species of this large Family, excepting those of the Sub-type represented by the Munjeet with whorled leaves, are almost exclusively tropical, abounding both in the Old and New World. Many of the commonest tropical weeds belong to the Family, as well as several beautiful garden flowers. Comparatively few species are of economic importance, but amongst these are several of very great value, as the Coffee-tree (*Coffea arabica*) and the Peruvian Barks (*Cinchona*), both now cultivated to profit in India ; the former principally in Ceylon, the latter on cool mountain slopes in Southern India and the Himalaya.

OBSERVE the invariably opposite leaves of the Family in all the genera, save one growing in the Himalaya and Khasia mountains (*Carlemannia*), with entire margins : the scale-like stipule between the bases of the opposite leaves, hence called *interpetiolar :* the recurved axillary spines of Gambir (*Uncaria Gambir*) : the connate capitate flowers of *Morinda :* the development of one calyx-lobe of the flower into a leaf in *Mussænda :* the horny albumen of the seeds, well seen in Coffee.

The principal supply of Coffee for the British market is obtained from Ceylon. In 1868 the quantity imported from that island amounted to one hundred millions of pounds ; the total from all sources, Ceylon included, to upwards of one hundred and seventy-three million pounds.

The Peruvian Barks now successfully introduced into India by the British Government, and into Java by the Dutch, are native on the slopes of the Peruvian Andes, where they are becoming scarcer and more difficult of access year by year, owing to the reckless way in which they are destroyed by the collectors of the medicinal bark.

The valuable emetic, Ipecacuanha, is the root of a Brazilian species of the Family (*Cephaëlis*). Several Indian Rubiaceæ are used in native medicine.

Gambir is an astringent extract, obtained by boiling down

the herbage of *Uncaria Gambir* in the Malay Islands and Peninsula.   It is exported for the use of tanners and dyers.

The Munjeet is a near ally of the Madder, cultivated in Europe and the Levant (*Rubia tinctoria*).   It affords the valuable dye called " Indian red."   The leaves of *Canthium parviflorum*, used as a fence-shrub, are eaten in curries.

52. Natural Order, *Compositæ.*—The Composite Family.

Trees, shrubs, or herbs, with alternate or opposite leaves and capitate inflorescence.   Anthers syngenesious.   Ovary one-celled, with one erect ovule.

TYPE—The Garden Zinnia (*Zinnia elegans*).

FIG. 141. *Zinnia elegans*, considerably reduced.

A common garden annual (imported from America), with opposite entire leaves, and terminal, solitary, showy heads of variously coloured florets.

Organ.	No.	Cohesion.	Adhesion.
Calyx. *sepals.*	5	Gamosepalous.	Superior.
Corolla. *petals.*	5	Gamopetalous.	Epigynous.
Stamens.	5	Syngenesious.	Epipetalous.
Pistil. *carpels.*	2	Syncarpous.	Inferior.
Seed solitary, erect, exalbuminous.			

N.B. The ovary is invariably one-celled, with a solitary erect ovule. The number of carpels forming the pistil is inferred from the bifid stigma. From analogy it is inferred also that the calyx consists of five sepals. The limb, however, is frequently wholly absent (*obsolete*), as in the Umbellate Family; while in some genera it is represented by scales or hairs, forming a *pappus*, which persists and crowns the ripe fruit.

In *Zinnia* the pappus is reduced to one or two bristles.

COMPARE, with respect to the form of the corolla of the florets of the flower-heads—

Zinnia, with disk and ray (see p. 36); *Vernonia cinerea* (an annual weedy Flea-bane, with small flower-heads, common in waste places), with all the florets regular (*tubular*), resembling those of the disk of *Zinnia* in form; *Sonchus* (a weed of cultivated ground, with yellow flower-heads), with all the florets irregular (*ligulate*), resembling those of the ray of *Zinnia* in form; *Echinops echinatus* (with terminal, globose, spiny flower-heads), each floret enclosed in a distinct involucel.

When all the florets of a flower-head are tubular, the flower-head is *discoid;* if the florets of the ray be ligulate, the flower-head is *radiate*.

The florets of the same flower-head may all be perfect
(*Vernonia, Sonchus*); or those of the disk may be perfect and
those of the ray pistillate (*Zinnia*), or the florets monœcious
(*Calendula*); or the outer florets may be neuter, or all the
florets of the flower-head may be unisexual and the heads
diœcious, as in the common Thistle of waste places (*Carduus
arvensis*).

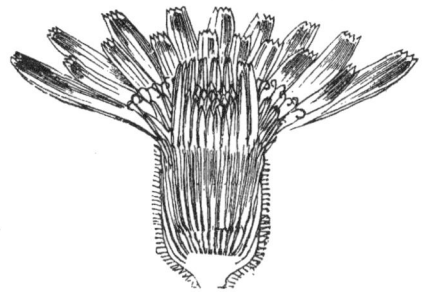

FIG. 142. Vertical section through flower-head of Sonchus.

If all the florets of a flower-head be perfect, it is said to
be *homogamous;* if some of them be imperfect, it is *heterogamous.*

Compare, also, the *common receptacle* of *Vernonia*, or of
any common *Blumea,* with that of *Zinnia.* The latter has
numerous chaffy scales, or minute bracts, called *pales,*
amongst the florets, and the common receptacle is said to
be *paleaceous;* the former are destitute of scales, and the
common receptacle is said to be *naked.*

The Composite Family is the largest in the vegetable
kingdom, and at the same time one of the best-defined and
most easily recognised. When once it is thoroughly understood that the " flower-heads " are not simple flowers, but
heads of (usually numerous) independent florets enclosed

within a whorl of bract-leaves (*involucre*), there remains no difficulty in comprehending the principal technical marks of the Family. The Composites are distinguished from allied families by—

1. Their inflorescence in flower-heads.

2. Their syngenesious anthers.

3. Their inferior one-celled ovary, with a single erect ovule. The Family is divided into three Tribes :—

1. With the perfect or hermaphrodite florets regular (*Tubulifloræ*).

2. With the florets bilabiate (*Labiatæfloræ*, principally South American).

3. With all the florets ligulate (*Ligulifloræ* or *Cichoraceæ*).

Sub-tribes are based upon the form of the style and of its divisions.

The Family is represented in every quarter of the globe, and in all zones : in the north temperate zone generally by herbaceous or shrubby species ; in the tropics by large trees, as well as herbs.

Notwithstanding the enormous number of species in the Family, reckoned at from 8,000 to 10,000, the proportion of which any general use is made by man is comparatively small. Many local species are used in medicine ; many are bitter and aromatic, abounding in an etherial oil ; and a considerable number are used as salad or pot-herbs in temperate countries—amongst the rest, Lettuce (*Lactuca*) and Artichoke (*Cynara*). Jerusalem Artichokes are the tubers of a *Helianthus*, allied to the Sunflower (*H. annuus*). These are both American herbs, common everywhere in Indian gardens : the seeds of the latter are edible, and yield a valuable oil. Safflower (*Carthamus tinctorius*) is cultivated to some extent in India for its flowers, which yield a rose-coloured dye.

53. Natural Order, *Campanulaceæ.*—The Bell-flower Family.

Herbs or shrubs, with alternate or opposite exstipulate leaves and milky juice.   Ovary two- three- (or more-) celled. Seeds indefinite.

TYPE— Wahlenbergia agrestis.

A common, spreading, annual weed, with alternate simple leaves, and solitary, terminal, regular, bell-shaped, white flowers.

Organ.	No.	Cohesion.	Adhesion.
Calyx. sepals.	5	Gamosepalous.	Superior.
Corolla. petals.	5	Gamopetalous.	Epigynous.
Stamens.	5	Pentandrous.	Epigynous.
Pistil. carpels.	3	Syncarpous.	Inferior.

Lobelia trigyna, an annual herb, with similar habit, and axillary blue flowers on slender peduncles, represents a distinct sub-type, differing in having an irregular corolla and the anthers syngenesious.   From Composites, which have also syngenesious anthers, the three-celled ovary and many seeds easily distinguish this sub-type.

Several Bell-flowers (*Campanula*) and Lobelias are favourite garden-plants in temperate countries.   Two small-flowered Campanulas of Northern India exhibit the curious phenomenon of dimorphic flowers.   Besides the flower of usual form, there are smaller ones, about the size of coriander-seeds, which never open, but which nevertheless mature abundant seed.   The latter must necessarily be self-fertilised, while the flowers of ordinary form are often, if not always, crossed by the pollen of other flowers of the same species. The part which these hermetically sealed flowers play in the economy of plants is not yet well made out. They are found in isolated species and genera belonging to widely different Natural Orders, and it would be worth while to look for fresh examples amongst Indian weeds.

54. Natural Order, *Ericaceæ.*—The Heath Family.

Shrubs or trees, with alternate simple leaves. Stamens usually eight or ten, with anthers opening by pores. Seeds indefinite. (In the tribe Ericeæ the ovary is free.)

TYPE—*Rhododendron arboreum* (or any other species of Rhododendron).

A mountain shrub, with evergreen, simple, coriaceous leaves, and terminal, umbellate clusters of red or white, slightly irregular, bell-shaped, beautiful flowers.

Organ.	No.	Cohesion	Adhesion.
Calyx. *sepals.*	5	Gamosepalous.	Inferior.
Corolla. *petals.*	5	Gamopetalous.	Hypogynous.
Stamens.	10	Decandrous.	Hypogynous.
Pistil. *carpels.*	10	Syncarpous.	Superior.

A Family widely spread, but most prevalent in cool climates, either of the temperate zones, or on mountains between the tropics. The Himalaya are especially rich in beautiful species of the genus *Rhododendron,* including the Type species, which is also found in the mountains of Southern India.

The Cranberry Tribe (*Vacciniaceæ*) constitute a sub-type, differing from the above in having the ovary inferior, and the stamens and corolla epigynous in consequence. Several species of this group occur in the mountains of Northern India.

OBSERVE the *declinate* (turned to one side) stamens, with the anthers opening by minute terminal pores, in *Rhododendron*. The anthers of most genera of the Family are curiously appendaged with bristle-like spurs.

But few species are turned to account by man, excepting for garden and shrubbery ornamentation. The genus Heath (*Erica*), a great favourite on account of the elegance and beauty of its flowers, is remarkably numerous in *species* at the Cape of Good Hope, while two or three social species, numerous in *individuals*, cover large areas of barren land in Northern Europe. The so-called *Azalea indica*, a favourite greenhouse shrub in Europe, is a Chinese plant. It differs from *Rhododendron* in having deciduous leaves.

** *Ovary superior.*

55. Natural Order, *Styraceæ*.—The Benzoin Family.

Trees or shrubs, with alternate simple leaves. Stamens often indefinite, inserted on base of the corolla-tube. Ovary sometimes free. Seed usually solitary.

TYPE—*Symplocos racemosa.*

Organ.	No.	Cohesion.	Adhesion.
Calyx. sepals.	5	Gamosepalous.	Half-inferior.
Corolla. petals.	5	Gamopetalous.	Perigynous.
Stamens.	∞	Polyandrous.	Epipetalous.
Pistil. carpels.	3	Syncarpous.	Half-superior.

A small tree, with alternate, smooth, serrulate leaves, and axillary and terminal racemes of yellow flowers.

A small, widely spread Family, including but few species of economic value. Some species of *Symplocos* are used as yellow dyes, and the bark of our Type is employed as a mordant in dyeing with Munjeet.

The fragrant resin Benzoin is the exudation of a Malayan *Styrax*.

56. Natural Order, *Ebenaceæ.*—The Ebony Family.

Trees or shrubs, with alternate entire leaves, and diœcious regular flowers. Stamens indefinite.

TYPE—*Diospyros cordifolia.*

A tree, with the older branches spinose, alternate simple leaves, and axillary, drooping, diœcious flowers.

The male flowers in clusters of about three, the female flowers solitary.

Organ.	No.	Cohesion.	Adhesion.
Calyx. *sepals.*	4	Gamosepalous.	Inferior.
Corolla. *petals.*	4	Gamopetalous.	Hypogynous.
♂ Stamens.	16	Polyandrous.	Epipetalous.
♀ Pistil. *carpels.*	8	Syncarpous.	Superior.

N.B. The stamens are very nearly hypogynous, being inserted upon the base of the corolla.

The genus Ebony (*Diospyros*) is the largest and most important of this rather small Family, which is chiefly confined to hot countries.

Several Indian species of Ebony afford valuable cabinet-wood, especially the true Ebony (*D. Ebenus*), and the Calamander (*D. quæsiti*) of Ceylon. Ebony is remarkable for the dark colour and gravity of its heart-wood. The sap-wood (*alburnum*) is pale or nearly white. A few species yield an edible fruit.

57. Natural Order, *Sapotaceæ.*—The Sapodilla Family.

Trees or shrubs, with alternate entire leaves. Flowers regular. Stamens opposite lobes of corolla, equal or twice as many, often with numerous scale-like staminodia.

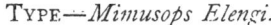

TYPE—*Mimusops Elengi.*

FIG. 143. *Mimusops Elengi* (called Bokul or Bukul), much reduced, with a detached fruit.

A much-branched, ornamental tree, common in gardens, with alternate, entire, smooth leaves, axillary fascicles of six to eight fragrant, white flowers, and yellow, edible berries.

Organ.	No.	Cohesion.	Adhesion.
Calyx. *sepals.*	8	Gamosepalous.	Inferior.
Corolla. *petals.*	24	Gamopetalous.	Hypogynous.
Stamens.	8	Octandrous.	Epipetalous.
Pistil. *carpels.*	8	Syncarpous.	Superior.

OBSERVE the unusual number of sepals and petals, due to the former occurring in two and the latter in several alternating series. In other genera the number of corolla-lobes is usually smaller. Observe, also, the so-called "barren stamens," or scales, alternating with the anther-bearing stamens.

Several species are cultivated in the tropics for the sake of their useful fruit; amongst the rest, the Sapodilla Plum (*Achras Sapota*) and Star Apple (*Chrysophyllum Cainito*). The kernels of the seeds of some species of *Bassia* contain much oil, or so-called "vegetable butter," which is collected from two or three species for use as food, for burning, and in soap-making. The dried flowers, also, are sold in bazaars in India.

Gutta Percha is the inspissated milk-sap of species of *Isonandra*, principally *I. Gutta*, a Malayan tree.

58. Natural Order, *Oleaceæ.*—The Olive Family.

Trees or shrubs with opposite leaves. Flowers regular. Stamens two.

Type—*Ligustrum robustum.*

A tree with opposite, entire, smooth leaves and terminal panicles of numerous, small, white flowers, resembling those of the European Privet.

Organ.	No.	Cohesion.	Adhesion.
Ca'yx. *sepals.*	4	Gamosepalous.	Inferior.
Corolla. *petals.*	4	Gamopetalous.	Hypogynous.
Stamens.	2	Diandrous.	Epipetalous.
Pistil. *carpels.*	2	Syncarpous.	Superior.

The Jessamines (*Jasmineæ*) form a distinct Tribe of this Family. *Jasminum Sambac,* a shrubby twiner, with simple leaves and delightfully fragrant (often double) flowers, and *J. grandiflorum,* or the common white Jessamine (*J. officinale*) of gardens, with pinnate leaves, may be taken as representatives of the Sub-type. They differ from the Type in having the corolla with five or more lobes, and a two-lobed fruit. To this Tribe belongs also the honey-scented, night-flowering *Nyctanthes arbor-tristis,* a small tree much cultivated in Indian gardens.

The two stamens distinguish Oleaceæ from nearly all gamopetalous Families with regular flowers.

To this small Family belong the Ash (*Fraxinus*), which is sometimes apetalous, and the Olive (*Olea europæa*); the latter originally native of Syria and Greece, but from a remote period extensively grown around the Mediterranean, for the sake of the excellent oil obtained from the pulp of its drupes.

59. Natural Order, *Apocynaceæ.*—The Dogbane Family.

Trees or shrubs, often climbers, with opposite (rarely alternate) entire leaves. Flowers regular. Stamens as many as, and alternate with, corolla-lobes. Carpels two, usually free below.

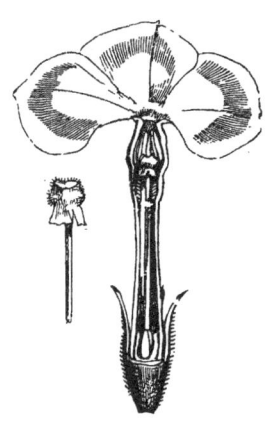

FIG. 144. *Vinca rosea*, one-third or one-half the natural size.

FIG. 145. Vertical section of flower of *Vinca rosea.* The transversely constricted annulate stigma to the left.

TYPE—*Vinca rosea.*

Organ.	No.	Cohesion.	Adhesion.
Calyx.. sepals.	5	Gamosepalous.	Inferior.
Corolla. petals.	5	Gamopetalous.	Hypogynous.
Stamens.	5	Pentandrous.	Epipetalous.
Pistil. carpels.	2	Syncarpous.	Superior.

A small erect perennial, everywhere in gardens, with opposite, entire, obtuse leaves, and pretty, axillary, rose or white flowers in pairs.

OBSERVE the twisted æstivation of the corolla-lobes, and their slight obliquity, common in the Family : the hour-glass narrowing of the stigma : the carpels, with the ovaries free, but the styles coherent, an unusual condition. The reverse condition, with coherent ovaries and free styles, is much more prevalent in syncarpous flowers. A pair of small glands alternate with the carpels upon the receptacle.

Compare with the fruit of 'Vinca, consisting of two fol-licles, that of *Carissa Carandas* (a very spinous shrub, serving for hedges), a two-celled berry used in tarts, pre-serves, &c.; and of *Allamanda*, a Brazilian climber, common in gardens, with large yellow flowers, and a one-celled fruit with parietal seeds.

The Dogbanes constitute a large Family, principally tropical, and many of them with very gay flowers. They abound in a milky juice, which is often poisonous. It is collected from a few species for the sake of the caoutchouc which it contains. The Madagascar Ordeal-tree (*Tan-ghinia*) belongs to this Family. It is said that the kernel of a single fruit suffices to poison twenty people. The Oleander (*Nerium*), of which one species is grown in India, is also poisonous. Many species afford useful medicines.

60. Natural Order, *Asclepiadaceæ.*—The Asclepias Family.

Shrubs or herbs, often climbing, with opposite entire leaves. Flowers regular. Anthers coherent ; pollen-masses adhering to the stigma. Carpels two, usually free below.

TYPE—The Mudar (*Calotropis gigantea*).

One of the commonest shrubs of India, abounding in milky juice ; erect, smooth, with opposite entire leaves, and

FIG. 146. Mudar (*Calotropis gigantea*), reduced   A ripe follicle to the
left below.

interpetiolar (not axillary, but from between the petioles on one side of the stem) umbels of pretty rose and purple flowers.

Organ.	No.	Cohesion.	Adhesion.
Calyx. sepals.	5	Gamosepalous.	Inferior.
Corolla. petals.	5	Gamopetalous.	Hypogynous.
Stamens.	5	Pentandrous.	Gynandrous.
Pistil. carpels.	2	Syncarpous.	Superior.

A Family of peculiar botanical interest, on account of the remarkable structure of the essential organs of the flower.

OBSERVE the stamens cohering around the pistil and the five lobes, alternating with and exterior to the stamens, forming what is termed the *corona :* the pollen of each anther-cell cohering into a single mass (*pollen-mass*), the pollen-masses of the adjoining cells of distinct anthers united in pairs to the stigma : in a few Indian genera (*Secamone, Toxocarpus*) the anthers are four-celled, so that there are twenty pollen-masses, four to each anther : the carpels, as in the Dogbanes, with free ovaries cohering to form a single style and stigma : the silky *coma* of the ripe seeds, contained in follicular fruits.

The milk of the Mudar is in repute in native medicine, and the fibre of its bark is valuable for textile purposes. *Hemidesmus* affords the so-called Indian Sarsaparilla, and a *Marsdenia* a very tenacious fibre. Several species are favourite garden flowers, especially the *Stephanotus* of Madagascar, and the species of *Hoya.* In the Indian and Australian climbing and creeping genus *Dischidia* the leaves are sometimes converted into pitchers.

61. Natural Order, *Loganiaceæ.*—The Strychnos Family.

Trees, shrubs, or herbs, with opposite entire leaves, often with interpetiolar stipules. Stamens alternate with corolla-lobes. Ovary two- (or more-) celled, with axile placentas.

TYPE—Nux-vomica (*Strychnos nux-vomica*).

A tree with opposite, broadly ovate, entire, three- to five-nerved leaves, interpetiolar stipules, small, rounded, terminal, cymose panicles of greenish-white flowers, and orange-yellow fruits.

Organ.	No.	Cohesion.	Adhesion.
Calyx. *sepals.*	5	Gamosepalous.	Inferior.
Corolla. *petals.*	5	Gamopetalous.	Hypogynous.
Stamens.	5	Pentandrous.	Epipetalous.
Pistil. *carpels.*	2	Syncarpous.	Superior.

Many genera of this Family may be regarded as Rubiaceæ with a superior ovary. Like that Family they are usually provided with interpetiolar stipules.

Some of the species differ very much in habit from *Strychnos;* but as none of them fill an important part in Indian vegetation, it is not needful to refer to them here. Several of the Family are virulently poisonous ; few more so than the Nux-vomica itself, from the seeds of which the alkaloid Strychnia is prepared. The seeds of a close ally, *S. potatorum*, are used in India as "clearing-nuts," causing the impurities of water to settle to the bottom when poured into vessels which have been rubbed out with one of the seeds.

62. Natural Order, *Gentianaceæ.*—The Gentian Family.

Herbs with opposite entire leaves and of bitter taste. Flowers regular. Ovary usually one-celled, with two parietal placentas.

TYPE—*Exacum tetragonum.*

An erect, annual herb, growing in grassy places, with opposite, five-nerved, entire, smooth leaves, and terminal cymes of showy blue flowers.

Organ.	No.	Cohesion.	Adhesion.
Calyx. sepals.	4	Gamosepalous.	Inferior.
Corolla. petals.	4	Gamopetalous.	Hypogynous.
Stamens.	4	Tetrandrous.	Epipetalous.
Pistil. carpels.	2	Syncarpous.	Superior.

N.B. The ovary is normally one-celled in this Family, with two parietal placentas, but in *Exacum* the placentas project into the cavity and meet, so as to make it almost completely two-celled.

A cool-climate Family, with several tropical representatives. The species are generally herbaceous, and nearly all of them are bitter, some of them extremely so, possessing valuable tonic properties. Our Type species is employed in India as a febrifuge.

The flowers of the Gentians are very gay, often a brilliant blue, yellow, or red. A few are aquatic; *Villarsia indica,* with yellow and white, bearded flowers, is frequent in lakes and tanks.

R 2

63. Natural Order, *Bignoniaceæ.*—The Bignonia Family.

Trees or climbing shrubs, with opposite, usually compound, exstipulate leaves. Flowers irregular. Stamens usually fewer than corolla-lobes. Ovary two-celled. Seeds numerous, winged.

FIG. 147. *Bignonia indica;* a capsule to the left : much reducea.

TYPE—*Bignonia (Calosanthes) indica.*

A large tree, with large, opposite, pinnately divided leaves (four to six feet long), terminal racemose panicles of trumpet-shaped, red or purple, slightly two-lipped flowers, and very long, flattened, two-valved capsules.

Organ.	No.	Cohesion.	Adhesion.
Calyx. *sepals.*	5	Gamosepalous.	Inferior.
Corolla. *petals.*	5	Gamopetalous.	Hypogynous.
Stamens.	5	Pentandrous.	Epipetalous.
Pistil. *carpels.*	2	Syncarpous.	Superior.
Seeds indefinite, broadly winged, exalbuminous.			

OBSERVE two stamens longer than the rest, indicating an approach to the didynamous condition, common in several Families with bilabiate gamopetalous corollas.

Many species are climbers in tropical countries both of the Old and New World. They are generally characterised by very showy flowers, and several of them are garden favourites. The West Indian Calabash-tree (*Cresentia*), with gourd-like fruits, the hard rind of which is made into various utensils, belongs to a tribe of this Family.

64. Natural Order, *Pedaliaceæ.*—The Sesamum Family.

Herbs with opposite leaves. Flowers irregular. Stamens fewer than corolla-lobes. Ovary with twice as many cells as carpels.

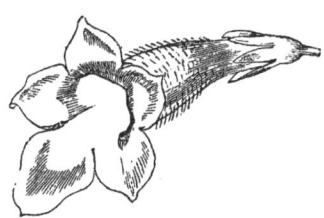

FIG. 148. Flower of *Sesamum indicum.*

TYPE—The Til, or Gingeley Plant (*Sesamum indicum*).

An annual pubescent herb, everywhere cultivated, two to four feet tall, with entire or three-lobed leaves and axillary, slightly bilabiate, pale rose-coloured flowers, forming terminal racemes.

Organ.	No.	Cohesion.	Adhesion.
Calyx. *sepals.*	5	Gamosepalous.	Inferior.
Corolla. *petals.*	5	Gamopetalous.	Hypogynous.
Stamens.	4	Didynamous.	Epipetalous.
Pistil. *carpels.*	2	Syncarpous.	Superior.

This small Family may almost be regarded as a Tribe of the foregoing.

OBSERVE the rudiment of the fifth stamen : the ovary spuriously four-celled, owing to the margins of the carpels being recurved so as to meet the dorsal sutures. The Type of the Family is the familiar Gingeley-oil plant, cultivated very extensively in warm countries for the sake of the oil expressed from its seeds.

Near allies of the Sesamums are the *Cyrtandreæ*, a tribe of the Family *Gesneraceæ*, including many herbs with very pretty flowers of the mountain valleys of India and of the Malayan islands. The species are so local that it is difficult to select a serviceable type. Some of the more beautiful Indian species belong to the genera *Æschynanthus* and *Didymocarpus*. They have usually tailed seeds, contained in long, straight or spirally twisted, very slender, siliquiform capsules. The beautiful Gloxinia and Achimenes of gardens are tropical American members of the Family.

65. Natural Order, *Convolvulaceæ.*—The Bindweed Family.

Herbs, shrubs, or rarely trees, usually twining or prostrate (*Cuscuta* parasitical), with alternate leaves and usually showy regular flowers with a plaited corolla. Ovary two- to four-celled. Seeds definite.

TYPE—The Elephant Creeper (*Argyreia speciosa*).

A twining shrub, with alternate, heart-shaped leaves silvery below with silky hairs, and large, handsome, axillary, rose-coloured, regular flowers.

Organ.	*No.*	*Cohesion.*	*Adhesion.*
Calyx. *sepals.*	5	Polysepalous.	Inferior.
Corolla. *petals.*	5	Gamopetalous.	Hypogynous.
Stamens.	5	Pentandrous.	Epipetalous.
Pistil. *carpels.*	2	Syncarpous.	Superior.
Seeds four in each fruit, cotyledons plicate.			

OBSERVE the membranous, accrescent bracts of the Indian genus *Neuropeltis :* the plaited æstivation of the corolla. Compare the berried fruit of *Argyreia* with the capsular fruit of any of the numerous Indian species of *Ipomœa* (two-celled), of *Pharbitis nil* (three-celled), or of the small-flowered *Poranas* (one-celled).

The Family is principally, but not wholly, tropical ; a few species occur in Britain, and one (*Convolvulus arvensis*), a common weed there, also occurs (introduced) in Northern India. The flowers of most of the genera are very handsome, and many of them are favourite garden climbers.

The Sweet Potato (*Batatas edulis*) is cultivated everywhere in tropical countries. It is the most important food-producing species of the Family. Its native country is uncertain, but evidence favours its American origin. Many of the Bindweeds are useful in medicine. A Mexican species (*Exogonium Purga*) affords the drug Jalap, and a *Convolvulus* of the Levant (*C. Scammonia*) Scammony.

The Dodders (*Cuscutæ*) constitute a remarkable Sub-type of this Family. They are twining, leafless, parasitical herbs, with small white corollas, often with minute scales inside, alternating with the stamens. Their embryo is spirally twisted, and is usually destitute of cotyledons. It germinates independently in the soil, but as early as possible lays hold of some neighbouring plant, to which it attaches itself by small suckers given off by its stem. One species, *C. reflexa*, is common in India.

66. Natural Order, *Boraginaceæ.*—The Borage Family.

Herbs, shrubs, or rarely trees, with alternate, simple, often rough leaves, and unilateral cymose inflorescence. Flowers regular, with stamens alternate with corolla-lobes. Ovary, four- (rarely two-) lobed, four- (two-) seeded.

Organ.	*No.*	*Cohesion.*	*Adhesion.*
Calyx. *sepals.*	5	Gamosepalous.	Inferior.
Corolla. *petals.*	5	Gamopetalous.	Hypogynous.
Stamens.	5	Pentandrous.	Epipetalous.
Pistil. *carpels.*	2	Syncarpous.	Superior.
Seeds, one in each cell of the four-lobed ovary.			

TYPE—*Trichodesma indicum.*

A spreading, scabrous or hairy annual, with entire, clasping leaves, and pale-blue regular flowers opposite to the bracts.

The Sweet-scented Heliotrope (*Heliotropium peruvianum*), common in gardens, may be taken as a Sub-type, differing from true Borages in its entire ovary and terminal style.

OBSERVE the characteristic roughness and harshness of the herbage of the Family: the one-sided racemose inflorescence, well shown in Heliotrope. It is made up of a series of distinct axes, each of which terminates in a single flower, while below it forms part of the common axis of the inflorescence. This explains the anomaly of the pedicels being opposite to the bracts, instead of axillary. Observe, also, the four-lobed ovary, composed of two carpels, the dorsal sutures of which are infolded so far as to divide each carpel into two one-seeded cells. This structure of the ovary closely resembles that of Labiates, from which Family the usually regular flowers and stamens equal in number to the corolla-lobes distinguish it, besides the usually alternate leaves and round stem of the Borages. To this Family belongs the Forget-me-not (*Myosotis*), common in wet places in Europe.

Regarded by some botanists as a distinct Family is the small group Cordiaceæ, represented by *Cordia Myxa*, a low tree with simple, alternate, entire or toothed leaves, small, white, panicled, polygamous flowers, and berried fruits. It differs from the Borages in having a twice-forked terminal style, baccate fruit, and plaited cotyledons.

67. Natural Order, *Solanaceæ.*—The Nightshade Family.

Herbs, shrubs, or sometimes trees, with alternate leaves and often extra-axillary inflorescence. Flowers nearly or

quite regular. Stamens alternate with corolla-lobes. Ovary
two-celled. Seeds indefinite.

TYPE—*Solanum Jacquinii.*

FIG. 149. *Solanum Jacquinii,* one-third to one-half the natural size.

A very prickly, spreading or creeping biennial or perennial,
with alternate or geminate lobed or pinnatifid leaves,
extra-axillary peduncles bearing a few blue or purplish
flowers, and yellow or pale-coloured berries.

Organ.	No.	Cohesion.	Adhesion.
Calyx. *sepals.*	5	Gamosepalous.	Inferior.
Corolla. *petals.*	5	Gamopetalous.	Hypogynous.
Stamens.	5	Pentandrous.	Epipetalous.
Pistil. *carpels.*	2	Syncarpous.	Superior.
Seeds ∞, albuminous.			

OBSERVE the frequently geminate, though not strictly opposite, leaves of the Solanums : the extra-axillary peduncle, well shown in the Type : the accrescent calyx of *Physalis* (Cape Gooseberry): the connivent anthers, opening by pores at their tips : the four-celled prickly fruit of the Thorn Apple (*Datura*), a weed of waste places ; the dorsal suture of the carpels is inflected so far as to meet the placentas, so that the ovary becomes spuriously four-celled—*spuriously*, because the dissepiments are not all the inflected *margins* of carpellary leaves, by which alone the ovary is normally divided into distinct cells. Hence, whenever there are more cells than carpels, some of the dissepiments are necessarily *spurious* in this technical sense.

Several species of the very large, chiefly tropical, genus Solanum are common in India, and any of them will serve as type in lieu of the above, as the Egg-plant, Brinjal, or Aubergine (*S. Melongena*), an introduced garden species, or the tomentose *S. verbascifolia.*

This Family includes, besides many very valuable food-plants, many dangerous, narcotic poisons. This anomaly, of species apparently so diverse in their properties being included in the same Natural Order, is explicable when we find that even the most useful food-products of the Family require to be cooked before they are fitted for use. The most important food-producing species are the Potato (*Solanum esculentum*), a South American herb, cultivated for the sake of its tubers throughout temperate countries, the Aubergine, and the Tomato (*Lycopersicum*). The so-called Cape Gooseberry (*Physalis peruviana*) is very common in India. The pulpy berry, which is wholly concealed within the yellowish, persistent, and accrescent calyx, is edible. The scarlet or orange-yellow fruits of species of *Capsicum* are everywhere used as a condiment.

The narcotic and poisonous species include Tobacco (*Nicotiana*), Deadly Nightshade (*Atropa*), Henbane (*Hyoscyamus*), and many others.

68. Natural Order, *Scrophulariaceæ.*—The Figwort Family.

Usually herbs with opposite or alternate leaves, and irregular flowers. Stamens fewer than corolla-lobes. Ovary two-celled. Seeds indefinite.

Type—*Maurandya semperflorens.*

Organ.	No.	Cohesion.	Adhesion.
Calyx. *sepals.*	5	Gamosepalous.	Inferior.
Corolla. *petals.*	5	Gamopetalous.	Hypogynous.
Stamens.	4	Didynamous.	Epipetalous.
Pistil. *carpels.*	2	Syncarpous.	Superior.
Seeds ∞, albuminous.			

A climbing (garden) herb, with alternate, hastate leaves, and axillary, solitary, pedunculate, irregular, rose-coloured flowers. (Or *M. Barclayana*, with purple flowers. Both introduced American species, commonly cultivated.)

OBSERVE the one-spurred corolla of *Linaria*, sometimes becoming five-spurred and regular (*Peloria*), by the development of a spur to each petal : the limb of the corolla, quinquepartite and nearly regular (*Verbascum*), quadripartite (*Veronica*) ; bilabiate corolla and personate (Snapdragon, *Antirrhinum*); ringent (*Pedicularis*); the anthers with their cells frequently divergent below, or actually separated by the dilatation of the connective in several genera.

Although this very large Family includes a great many Indian species, I have selected as Type one which is not native, but likely to be generally accessible in gardens, because many of the native species are insignificant weeds and difficult to identify, or else locally distributed.

The Family approaches Solanaceæ very nearly, but it may generally be distinguished from that Order by the deficiency of one or three stamens, the stamens thus becoming fewer than the number of petals cohering to form the corolla. When the fifth stamen is present, as in *Verbascum*, characters afforded by the æstivation of the corolla-lobes are made use of technically to separate the Orders. The common Mullein (*Verbascum Thapsus*) occurs in waste places in India. Several genera exhibit a partial parasitism, owing to the roots attaching themselves to the roots of plants amongst which they grow. *Pedicularis*, of which a large number of species occurs in the Himalaya, and *Rhinanthus* are examples of this condition, which unfits them for cultivation, although many of them bear very gay flowers. *Aeginetia indica* or *Aeg. pedunculata* may serve as a Sub-type, representing the tribe of Broomrapes (*Orobanchaceæ*). They are scaly parasites, destitute of green leaves, with large purplish flowers and spathe-like calyxes. The ovary is one-celled, with two double and much-branched placentas, covered with ovules. They are often found on the roots of grasses ; *Aeg. pedunculata* on that of the Kus-kus (*Andropogon muricatus*).

Many gay garden-flowers belong to the Figwort Family, as *Calceolaria, Pentstemon, Paulownia, Mimulus*, and *Torenia*. It affords, however, but few plants of economic value, excepting as medicines.

69. Natural Order, *Lentibulariaceæ.*—The Butterwort Family.

Herbs growing in water or damp places. Flowers two-lipped. Stamens two. Ovary one-celled, with a free central placenta.

Type— *Utricularia stellaris.*

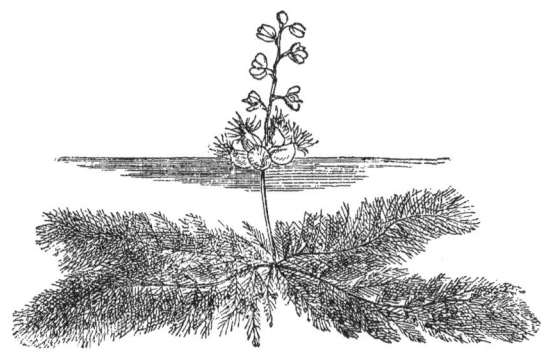

Fig. 150. *Utricularia stellaris.*

A floating plant, the submerged leaves with capillary segments bearing small bladders, and an erect peduncle with a whorl of inflated vesicles below, terminating in a raceme of irregular yellow flowers.

Organ.	No.	Cohesion.	Adhesion.
Calyx. *sepals.*	5	Gamosepalous.	Inferior.
Corolla. *petals.*	5	Gamopetalous.	Sub-hypogynous.
Stamens.	2	Diandrous.	Epipetalous.
Pistil. *carpels.*	2(5?)	Syncarpous.	Superior.
Seeds ∞, on free central placenta.			

A small but distinct Family of water and bog plants widely dispersed both in tropical and temperate countries. They are nearly allied to the Family last described, but differ in their one-celled ovary, with a free central placenta, similar to that of the Primroses, from which their irregular diandrous flowers distinguish them.

OBSERVE the minute "bladders" attached to the submerged leaves and radicular fibres of many species of Bladderwort (*Utricularia*). Several small Indian species, growing on the ground, are leafless at the time of flowering. *U. reticulata*, a species with large, beautiful purple flowers, is common in rice-fields. It is a variable plant in its habit and the size of its flowers. The larger forms of it are twining; the smaller, rigid and erect.

70. Natural Order, *Acanthaceæ.*—The Acanthus Family.

Usually herbs or shrubs, with opposite, simple leaves. Flowers irregular, usually bracteate. Ovary two-celled. Seeds usually supported on cushions or hooks of the placenta.

TYPE—*Justicia Adhatoda* (Bakas).

A small tree, with opposite, entire leaves, and axillary

bracteate spikes of bilabiate white flowers, spotted with red
or purple, in the axils of herbaceous bracts.

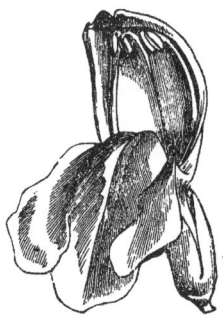

FIG. 151. Flower of *Justicia Adhatoda.*

Organ.	No.	Cohesion.	Adhesion.
Calyx. *sepals.*	5	Gamosepalous.	Inferior.
Corolla. *petals.*	5	Gamopetalous.	Hypogynous.
Stamens.	4	Didynamous.	Epipetalous.
Pistil. *carpels.*	2	Syncarpous.	Superior.

OBSERVE the two-celled, undivided ovary, by which the
Family may be distinguished from Labiates, which it resembles in its opposite leaves and often bilabiate flowers :
the capsule, dehiscing by two valves, each valve bearing the
seeds upon its median line, supported upon hard, hooked
processes.

Sub-type—*Thunbergia grandiflora*, a climbing, woody
perennial, with axillary, peduncled, large blue and white,
bibracteate flowers.

The Sub-type differs in its twining stem, in the very small calyx, reduced to a mere ring, and in the little cups supporting the seeds. The pair of valvate bracts must not be mistaken for a calyx.

This very large tropical Family includes many insignificant weeds, and many species with beautiful and showy flowers; very few, however, are turned to any economic use. A few are employed in native medicine, and one or two afford dyes.

71. Natural Order, *Labiatæ.*—The Labiate Family.

Herbs or shrubs, usually aromatic, with opposite leaves and irregular flowers. Stamens two or four. Ovary four-celled and deeply four-lobed.

TYPE—Sweet Basil (*Ocymum basilicum*).

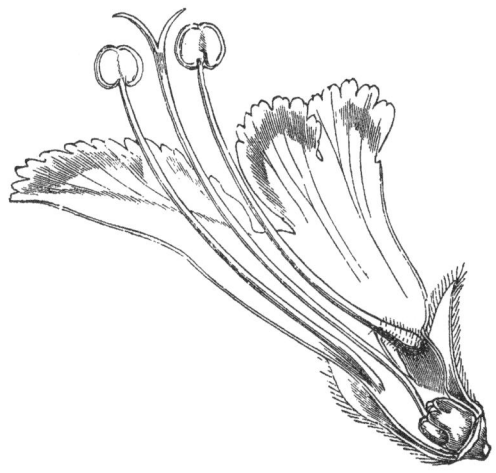

FIG. 152. Vertical section of flower of *Ocymum*, much enlarged.

S

An erect or ascending aromatic shrub, with square stem, opposite, simple leaves, and terminal racemes of white or pink, bilabiate flowers, arranged in six-flowered verticillasters.

Organ.	No.	Cohesion.	Adhesion.
Calyx. *sepals.*	5	Gamosepalous.	Inferior.
Corolla. *petals.*	5	Gamopetalous.	Hypogynous.
Stamens.	4	Didynamous.	Epipetalous.
Pistil. *carpels.*	2	Syncarpous.	Superior.
Seeds solitary, in each of the four lobes of the fruit.			

OBSERVE the square stem and invariably opposite leaves of the Family: the apparently whorled flowers of the Type, made up of a pair of opposite three-flowered cymes : the upper tooth of the calyx of Sweet Basil, which enlarges after flowering, becoming membranous, with decurrent adnate margins : the diandrous flowers of *Salvia*, with the anther-cells removed from each other by a long (*distractile*) connective, which must not be mistaken for the true filament, which is very short : the four-lobed ovary, similar to that of Borages (page 248), with the style rising from between the lobes (*gynobasic*).

A large, widely-distributed Family, characterised by aromatic properties. None of the species are hurtful, and many are useful pot-herbs, as Mint (*Mentha*), Marjoram (*Origanum*), Thyme (*Thymus*), Sage (*Salvia*), and our Type-species. The fragrant oils of some species, as Lavender (*Lavendula*) and Patchouli (*Pogostemon*), are in request as perfumes, while others are serviceable stimulant medicines.

72. Natural Order, *Verbenaceæ.*—The Verbena Family.

Trees, shrubs, or herbs, with opposite leaves and more or less irregular flowers.   Stamens usually four.   Ovary four-celled, entire.

TYPE—The (quinque-foliolate) Chaste-tree (*Vitex Negundo*).

A small, downy tree or shrub, with opposite tri- or quinque-foliolate leaves, and terminal panicles of small, purplish-blue, irregular flowers.

Organ.	No.	Cohesion.	Adhesion.
Calyx. *sepals.*	5	Gamosepalous.	Inferior.
Corolla. *petals.*	5	Gamopetalous.	Hypogynous.
Stamens.	4	Didynamous.	Epipetalous.
Pistil. *carpels.*	2	Syncarpous.	Superior.

OBSERVE the enlarged limb of the calyx, coloured bright scarlet, of *Holmskioldia :* the four-celled ovary of the Family, resembling that of Labiates, excepting that it is not lobed and the style is terminal.

A rather large Family, principally confined to tropical countries.   The species are generally trees or shrubs, though several are low herbs, as the Garden Verbena, one of the brightest bedding-plants of English gardens.   Teak (*Tectona grandis*) is the species of first importance in India, affording one of the best and most durable timbers known.   The flowers of the Teak are often hexandrous.   Normally they are pentandrous.

Aloysia citriodora, the Lemon-scented Verbena, a fragrant South American shrub, is common in Indian gardens, and several species of the genera *Lantana* and *Clerodendron* are sought after for the beauty of their flowers. The only

FIG. 153. Teak (*Tectona grandis*).   To the left the accrescent calyx, enclosing the fruit.

British representative of the Family, *Verbena officinalis*, is a wide-spread weed, with inconspicuous flowers, sufficiently common in India.

73. Natural Order, *Myrsinaceæ.*—The Ardisia Family.

Trees or shrubs, with alternate, simple leaves, and regular flowers. Stamens five or four, opposite lobes of corolla. Ovary one-celled, with free central placenta.

Type—*Ardisia humilis.*

A shrub, with alternate, entire, smooth leaves, axillary, short, umbelliform racemes of small, pale rose-coloured flowers, and one-seeded berries.

Organ.	*No.*	*Cohesion.*	*Adhesion.*
Calyx. *sepals.*	5	Gamosepalous.	Inferior.
Corolla. *petals.*	5	Gamopetalous.	Hypogynous.
Stamens.	5	Pentandrous.	Epipetalous.
Pistil. *carpels.*	5	Syncarpous.	Superior.

Observe the stamens, which are opposite to the lobes of the corolla: the free central placenta bearing numerous ovules, of which but one is perfected into a seed.

A Family tolerably abundant in hot countries, where it represents the Primrose Order, from which it has little technically to separate it, excepting that the Myrsineæ are, as a rule, shrubs or trees, with more or less succulent fruits, while the Primulaceæ are herbs, with dry, capsular fruits.

In some genera of this Family, native in India, the petals are free to the base; in *Ægiceras*, a tree growing on the coast, the anther-cells are transversely chambered, and in *Mæsa* the ovary is more or less adherent to the calyx. In other respects the essential characters of the Type-species apply to the rest of the Family.

Very few species are turned to economic account.

74. Natural Order, *Primulaceæ.*—The Primrose Family.

Herbs with regular flowers. Stamens opposite the lobes

of the corolla and equal in number.   Ovary one-celled, with a free central placenta.

TYPE—*Primula sinensis* (or any species of garden or wild Primrose).

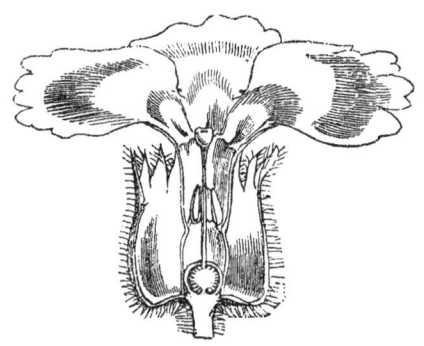

FIG. 154. Vertical section of flower of *Primula sinensis.*

Herb with radical leaves and umbellate or verticillate regular flowers.

Organ.	No.	Cohesion.	Adhesion.
Calyx. sepals.	5	Gamosepalous.	Inferior.
Corolla. petals.	5	Gamopetalous.	Hypogynous.
Stamens.	5	Pentandrous.	Epipetalous.
Pistil. carpels.	5	Syncarpous.	Superior.

OBSERVE the stamens opposite to the corolla-lobes, and the free central placenta, as in the foregoing Family.   Many species, especially of the genus Primrose, are prime spring favourites in the gardens of England.   The native Indian representatives of the Family are almost confined to the mountains, many of them attaining an alpine elevation. The Pimpernel (*Anagallis arvensis*), a common English corn-field weed, with small bright red or blue flowers, occurs in India.   It is remarkable in the mode of dehiscence of its capsule, which opens transversely, the upper part separating like a cap.

75.   Natural Order, *Plumbaginaceæ.*—The Thrift Family.

FIG. 155. *Plumbago zeylanica*, reduced.

Herbs or shrubs.   Petals and stamens five, nearly or quite free.   Ovary one-celled, with five styles.

TYPE— *Plumbago zeylanica.*

A shrubby garden plant, with long branches, alternate,

simple leaves, and long, viscid-hairy spikes of regular white flowers (or the Rose Plumbago, with rose-coloured, or Cape Plumbago with pale blue, flowers).

Organ.	No.	Cohesion.	Adhesion.
Calyx. sepals.	5	Gamosepalous.	Inferior.
Corolla. petals.	5	Gamopetalous.	Hypogynous.
Stamens.	5	Pentandrous.	Hypogynous.
Pistil. carpels.	5	Syncarpous.	Superior.

OBSERVE the hypogynous stamens, opposite to the corolla-lobes : the petals, in some species, nearly or quite free : the one-celled ovary with a solitary ovule, suspended from a slender cord (*funicle*) rising from the base of the cavity.

The three species named above are common ornamental species in Indian gardens. Of the other genera of this small Family, many species are sea-coast or salt-marsh plants, or natives of the sterile saline wastes to the north-west of India. Some species are used in native medicine.

76. Natural Order, *Plantaginaceæ.*—The Plantain Family.

Herbs with small greenish, often spicate, flowers, with scarious corollas. Stamens four, exserted. Style simple.

Organ.	No.	Cohesion.	Adhesion.
Calyx. sepals.	4	Polysepalous.	Inferior.
Corolla. petals.	4	Gamopetalous.	Hypogynous.
Stamens.	4	Tetrandrous.	Epipetalous.
Pistil. carpels.	2	Syncarpous.	Superior.

Type—Greater Plantain (*Plantago major*).

Herb of waste places, with broad radical leaves, and small, spicate, greenish flowers.

Observe the dry, scarious corolla : the slender filaments : the transverse dehiscence of the small, capsular fruit.

A small, widely-spread Family, with inconspicuous flowers, usually arranged in spikes. The species serving as Type is a common weed in Europe as well as in India. The flowers of the Indian form are often more loosely arranged upon the long spike than in the common English plant. The seeds of one species (*P. decumbens* [*Ispaghula*]) are sold in the bazaars : they are useful in medicine owing to their cooling mucilage.

Division—INCOMPLETÆ.

77. Natural Order, *Nyctagineæ.*—The Marvel-of-Peru Family.

Herbs or shrubs, with alternate or opposite, unequal leaves. Base of the perianth-tube persistent, enclosing the one-celled, one-seeded, superior ovary.

Type—*Boerhaavia diffusa.*

A common perennial, procumbent weed, with opposite, more or less unequal, ovate or cordate leaves, and loose panicles of small red or white flowers, collected in small heads.

Organ.	No.	Cohesion.	Adhesion.
Perianth. *leaves.*	5	Gamophyllous.	Inferior.
Stamens.	3	Monadelphous.	Hypogynous.
Pistil. *carpels.*	1 (?)	Apocarpous(?).	Superior.
Seeds solitary ; embryo curved round the albumen.			

Or, in lieu of the above, the Marvel of Peru (*Mirabilis*), frequent in gardens, with white, crimson, yellow, or variegated flowers.

OBSERVE the opposite leaves, one of each pair being generally smaller than the other: the five-leaved involucre of *Mirabilis*, which must not be mistaken for a calyx: the perianth contracted immediately above the ovary, to which it is not adherent. The upper portion of the perianth separates at the contraction after flowering, and the lower part of the perianth persists, becomes more or less hardened, and, enlarging, forms an outer envelope of the fruit. A similar condition occurs in the *Elæagnus* Family. One or two species of the South American genus *Bougainvillea* have large membranous, beautiful rose-coloured bracts, which render them favourite ornamental garden plants.

78. Natural Order, *Chenopodiaceæ.*—The Goosefoot Family.

Usually herbs with alternate or opposite leaves, and minute, herbaceous flowers. Ovary one-celled, uniovulate. with two or three stigmas.

TYPE—Common Goosefoot (*Chenopodium album*).

FIG. 156. Vertical section of flower of Goosefoot (*Chenopodium*).

An erect, annual, more or less mealy herb (two to eight feet high), with alternate leaves, and leafy, interrupted panicles of minute, clustered, greenish flowers.

Organ.	No.	Cohesion.	Adhesion.
Perianth. *leaves.*	5	Gamophyllous.	Inferior.
Stamens.	5	Pentandrous.	Hypogynous.
Pistil. *carpels.*	2	Syncarpous.	Superior.
Seeds solitary ; embryo curved round mealy albumen.			

OBSERVE the stamens opposite to the segments of the perianth (strictly, they are monadelphous, but the amount of cohesion is so slight as to be scarcely perceptible) : the one-celled ovary, horizontally flattened ; in the female flowers of the allied, weedy, polygamous genus *Atriplex*, the ovary is vertically compressed.

A widely-spread Family, occurring generally as weeds in waste places on the sea-shore, or in saline desert regions. Several peculiar genera abound in the Caspian and Aral region of Western Asia.

Several useful pot-herbs belong to the Family. The Type-species is thus employed in India. Beet (*Beta*) yields in cultivation a valuable tap-root; one variety of which, Mangold Wurzel, is grown for cattle in Europe. Another variety contains much saccharine juice, and is used in the manufacture of sugar on the continent of Europe. *Atriplex hortensis* is used as a pot-herb in the Deccan, as are species of *Basella* in India generally. The latter differs from the Goosefoots proper in its climbing habit and in the perianth, which at length becomes fleshy, forming a pseudo-pericarp, or baccate fruit. The perianth of *Basella* is in two series, an outer of two segments, an inner of five.

79. Natural Order, *Amaranthaceæ.*—The Amaranth Family.

Herbs (or shrubs, *Deeringia*), with opposite or alternate, simple, stipulate leaves, and minute scarious flowers.

TYPE—Cockscomb (*Celosia cristata*).

An erect annual, common in gardens, with alternate, simple leaves, and terminal, close, panicled or crested spikes of purple or golden, small, scarious flowers.

Organ.	*No.*	*Cohesion.*	*Adhesion.*
Perianth. *leaves.*	5	Polyphyllous.	Inferior.
Stamens.	5	Monadelphous.	Hypogynous.
Pistil. *carpels.*	2 (?)	Syncarpous.	Superior.
Seeds several ; embryo curved.			

OBSERVE the dry, scarious segments of the perianth, by which the Family may generally be distinguished at sight from the foregoing. The technical distinctions are not strong. A cultivated form of the Type is common (the Garden Cockscomb) in which the inflorescence is crested and laterally compressed, and many of the flowers are barren.

The Globe Amaranth (*Gomphrena*), Love-lies-bleeding and Princes' Feather (species of *Amaranthus*), are ornamental garden plants, while several weedy species, with insignificant, greenish flowers, serve as pot-herbs in India.

A common Indian plant, *Deeringia celosioides*, represents a Sub-type, differing from the Amaranths in climbing habit and baccate fruit.

80. Natural Order, *Polygonaceæ.*—The Buckwheat Family.

Herbs or shrubs, with alternate simple leaves, and sheathing stipules. Flowers very small, ovary one-celled, with three or two stigmas.

TYPE—*Polygonum barbatum.*

An erect, weedy herb, with alternate lanceolate leaves, sheathing, coarsely-bearded stipules, and long, terminal, spike-like racemes of inconspicuous, rose-coloured flowers.

Organ.	No.	Cohesion.	Adhesion.
Perianth. *leaves.*	5 (4)	Gamophyllous.	Inferior.
Stamens.	8 (5)	Oct-(pent-)androus.	Epiphyllous.
Pistil. *carpels.*	3	Syncarpous.	Superior.
Seeds solitary, erect, albuminous.			

OBSERVE the sheathing, membranous stipules characteristic of the Family : the three-cornered, one-celled ovary : trimerous symmetry of the flowers of Rhubarb (*Rheum*) and Dock (*Rumex*) : the three inner perianth-leaves enlarged after flowering and adpressed to the ovary.

A tolerably large and widely-spread Family. Many species are weeds of both tropical and temperate countries. A climbing *Polygonum*, the Buckwheat (*P. Fagopyrum*), is grown in the Himalaya for its farinaceous seeds, of which a good bread is made. The seeds of our Type are used in Hindoo medicine.

81. Natural Order, *Urticaceæ.*—The Nettle and Fig Family.

Trees, shrubs, or herbs, usually with alternate stipulate leaves and minute unisexual flowers. Stamens equal in number and opposite to perianth-lobes (except Sub-type 2). Ovary free, usually one-celled.

TYPE—Grass-cloth Plant (*Bœhmeria nivea*).

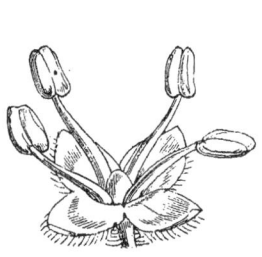 

FIG. 157. Staminate flower of Grass-cloth Nettle.　FIG. 158. Pistillate flower of same.

An erect shrub, with alternate, cordate leaves hoary beneath, and small, diœcious, greenish-yellow flowers, in axillary, peduncled, globose heads.

Organ.	No.	Cohesion.	Adhesion.
♂ Perianth. *leaves.*	4	Polyphyllous.	Inferior.
♀ Perianth. *leaves.*	4	Gamophyllous.	Inferior.
♂ Stamens.	4	Tetrandrous.	Hypogynous.
♀ Pistil. *carpels.*	2	Syncarpous.	Superior.
Seed solitary, albuminous.			

OBSERVE the tenacity of the liber of the bark, from which the so-called " grass-cloth" is prepared, and for the sake of which it is cultivated in Bengal : the stamens

FIG. 159. Longitudinal section of fruit of Nettle (*Urtica*).

FIG. 160. Section of seed of same, showing the large embryo, with a superior radicle and but little albumen.

opposite to the leaves of the perianth, with elastic filaments : the stinging hairs of the herbage of several Indian species (not species of *Bœhmeria*).

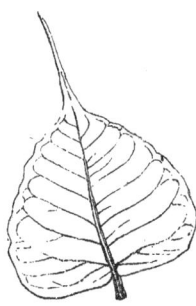
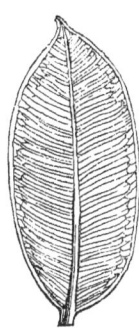

FIG. 161. Leaf of Peepul (*Ficus religiosa*).

FIG. 162. Leaf of India-rubber Fig (*F. elastica*).

FIG. 163. Leaf of Banyan Fig (*F. benghalensis*).

This large and very important Family embraces genera so diverse in the form of their inflorescence, that it is necessary to employ two or three Sub-types.

SUB-TYPE 1—Banyan, or Indian Fig (*Ficus benghalensis*), or the Peepul (*F. religiosa*). Flowers unisexual, in sessile, axillary, globular receptacles (Figs). Male and female flowers in the same Fig. Perianth three- to five-lobed : male flowers with a single stamen.

OBSERVE the adventitious roots given off freely by the branches (of the Banyan and many other species of *Ficus*), which descend and take root in the soil, thus enabling the tree to spread itself over a wide area. Observe, also, the large deciduous stipules, which leave ring-like scars at the base of the leaves when they fall. The stipules of some species (*F. elastica*, &c.) are well adapted for exhibiting the milk-sap canals *in situ* under the microscope. The margin of a young stipule should be placed in focus, and then pressure applied, or a wound made in some part of the stipule out of the field of view. This occasions a rapid movement or emptying of the fluid, apparently from the elasticity of the walls of the canals which contain it. Some botanists have thought that these canals formed a circulatory system analogous to the veins and arteries of animals, but this notion is not supported by careful observation. Note the succulent, hollow receptacle, the top of which is closed by minute scales. The flowers are very minute, and are closely packed on the inner surface of the common receptacle. When the fig is ripe, the individual fruits are commonly, but incorrectly, called the seeds. The common receptacle of Fig resembles the receptacle of Rose, with this important difference, that instead of enclosing the fruit-carpels of a single flower it encloses those of an indefinite number of flowers.

SUB-TYPE 2—Jack-fruit Tree (*Artocarpus integrifolia*).

A tree much cultivated in Southern India, with stipulate

leaves like the Figs, the flowers closely packed upon the outside of large oblong spikes, the male and female flowers on distinct spikes.   The perianth of the male is two-lobed, enclosing a single stamen ; of the female, entire and more

FIG. 164. Jack-fruit (*Artocarpus integrifolia*), much reduced.

or less adherent to adjoining perianths : so that the whole grow together, and when mature form a huge, collective fruit, ten to sixty pounds in weight.   The seeds are exalbuminous.

SUB-TYPE 3—Hemp (*Cannabis sativa* [Gunjah]).

A diœcious herb, largely cultivated for its narcotic, resinous leaves and flowers, which are smoked as " Bhang."

OBSERVE the digitate-partite leaves : the five-leaved perianth of the male flowers, and single folded sepal of the female : the resinous secretion of the younger parts, collected as the intoxicant " Churrus."

T

To the Fig Sub-type belongs the Fig (*F. Carica*), one of the few species of the large and generally tropical

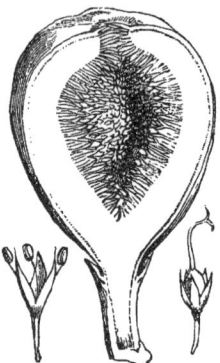

FIG. 165. Section of Fig (*Ficus Carica*), showing numerous flowers covering the inner surface of the hollow receptacle. To the left a detached staminate flower; to the right a pistillate flower.

genus *Ficus* affording an eatable fruit. It is supposed to have been native originally in Western Asia or in the

FIG. 166. Pistillate flower of Fig.

FIG. 167. Longitudinal section of achene and albuminous seed of Fig.

Mediterranean region, where it has been cultivated from a very remote period.

The species often yield a milky juice containing caout-chouc, especially *Ficus elastica*, from which Indian caoutchouc (India-rubber) is obtained. It is a fine indigenous tree, with glossy coriaceous leaves. The viscid juice of the Banyan is used to make birdlime.

A near ally of the Fig is the Mulberry (*Morus*), also with a collective, or multiple, fruit; but with the individual flowers arranged on the outside of a common receptacle, forming a short thick spike. The succulence of the fruit is due to the pulpy thickening of the leaves of the perianth. The tree is largely grown, both in Asia and Southern Europe, on account of its leaves, which are the food of the silkworm. The Paper Mulberry (*Broussonetia*) of Japan and Polynesia, the

Fig. 168. Collective fruit of Mulberry.

fibrous bark of which is beaten out and worked up in "Tapa-cloth," belongs to this group.

To the Jack-fruit Sub-type belong the Bread-fruit (*Artocarpus incisa*) of Polynesia and the dangerous Upas poison (*Antiaris toxicaria*) of Java. In Southern India the natives make "natural sacks" of the bark of *Antiaris saccidora*, which they separate from the trunk by beating it.

To the Hemp Sub-type belongs the Hop (*Humulus*), anomalous in the Family on account of its twining stem. The bracts of the female inflorescence enlarge after flowering, forming a loose cone. They are covered with microscopic glands containing a resin (*lupulin*), analogous to that secreted by Hemp (*Churrus*) in India, and possessing similar properties. Hence its use in Europe in malt liquors, to which it imparts flavour and a preserving quality.

T 2

82. Natural Order, *Euphorbiaceæ.*—The Spurge Family.

Trees, shrubs, or herbs, with unisexual flowers. Ovary free, usually three-celled, with one or two ovules in each cell.

TYPE—Castor-oil Plant (*Ricinus communis*).

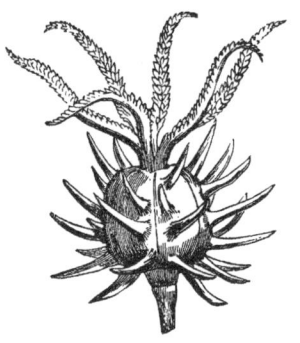
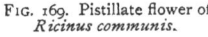

FIG. 169. Pistillate flower of *Ricinus communis*.

FIG. 170. Staminate flower of same.

An erect, smooth and glaucous, cultivated plant, sometimes tree-like ; with large alternate, peltate, palmately-lobed leaves, and terminal, racemose panicles of monœcious flowers ; the female flowers being above.

Organ.	No.	Cohesion.	Adhesion.
Perianth. *leaves.*	3-5	Gamosepalous.	Inferior.
Stamens.	∞	Polyadelphous.	Hypogynous.
Pistil. *carpels*	3	Syncarpous.	Superior.
Seeds one in each cell, albuminous.			

A very large Family, especially in tropical countries, where many species are turned to useful account.

Apart from the ovary, which is superior and usually three-celled with one or two ovules in each cell, the structure of the flowers is very diverse in the different tribes of this Family; so that it will be convenient to examine three Sub-types, comparing each of them with Castor-oil.

SUB-TYPE 1—Indian Spurge (*Euphorbia indica*).

An annual, ascending or decumbent weed, with milky juice, small opposite simple leaves, and minute greenish

FIG. 171. Involucre of Spurge, bordered by four horned "glands," and containing numerous male flowers and one female. The pedicel of the latter is curved over the side of the involucre.

flowers, in small axillary or terminal cymose heads.   Any other common species of *Euphorbia* will serve.

Organ.	No.	Cohesion.	Adhesion.
Perianth.	0	...	...
♂ Stamen.	1	Monandrous.	...
♀ Pistil. carpels.	3	Syncarpous.	...

OBSERVE the succulent, leafless, prickly *E. antiquorum*, with its three-angled irregularly narrowed stem, bearing the peduncles in the angles of the upper lobes: the bright scarlet pair of connate bracts, immediately under the flowers, in the introduced, showy garden species *E. splendens* and *E. Bojeri:* the achlamydeous flowers, of which a number of monandrous males (single stamens) and one female (pistil with three-celled ovary) are enclosed in a minute cup-like involucre, which beginners are apt to mistake for a calyx enclosing the stamens and pistil of a single flower. Careful examination with a magnifier of one of the stamens will show that the filament is jointed. This joint indicates the base of the monandrous flower, and is the point from which, in an allied species, the perianth is developed. The ovary is supported upon a stalk in the middle of the involucre, which stalk is jointed like the filaments. The involucre is usually bordered by four or five spreading lobes, which must not be mistaken for petals. These lobes are called the "glands" of the involucre.

SUB-TYPE 2—The Coral Plant (*Jatropha multifida*).

A succulent plant, common in gardens, with digitately multifid leaves, and small red flowers in cymes, borne upon erect, succulent peduncles.

Organ.	No.	Cohesion.	Adhesion.
Calyx. sepals.	5	Gamosepalous.	Inferior.
Corolla. petals.	5	Polypetalous.	Hypogynous (?).
♂ Stamens.	10	Monadelphous.	Hypogynous.
♀ Pistil. carpels.	3	Syncarpous.	Superior.

OBSERVE the presence of a corolla, unusual in the
Family. It occurs also in the Croton-oil Plant (*Croton
Tiglium*). This character, although exceptional in the

FIG. 172. Coral Plant (*Jatropha multifida*), reduced.

Family, does not, in accordance with principles stated
before, prevent its being classed with Incompletæ.

SUB-TYPE 3—*Phyllanthus Emblica.*

A tree with distichous, simple leaves, and small axillary,
monœcious flowers.

Organ.	No.	Cohesion.	Adhesion.
Perianth. *leaves.*	6	Polyphyllous.	Inferior.
♂ Stamens.	3(-5)	Monadelphous.	...
♀ Pistil. *carpels.*	3	Syncarpous.	Superior.

OBSERVE the leaves disposed along the twigs in two rows, so that the latter resemble long pinnate leaves: the two-ovuled cells of the ovary, as in Cicca, common in gardens, cultivated for the sake of its succulent fruit, which is eaten or made up in preserves.

A large number of species of the Spurge Family are dangerously poisonous, and many afford very valuable medicines, as the Castor- and Croton-oil Plants, referred to above. From the abundant deposit of starch in the roots of species of Manihot (allied to *Jatropha*), native in tropical South America, but widely cultivated in hot countries, they become most valuable food-plants, yielding Cassava-meal or Mandiocca, and Tapioca.

The flowers are usually so small, that the Family is not very serviceable for ornamental purposes. The bracts of the two species of *Euphorbia*, common in Indian gardens, named above, and especially of the Mexican *Poinsettia*, are, however, very brilliant, and compensate for the insignificance of the flowers.

South American India-rubber is obtained from the milk-sap of species of *Siphonia* (*S. brasiliensis*, &c.) The sap is dried upon clay bottle-shaped moulds, and when a sufficient thickness is obtained, the clay is washed out. The Kamala dye of India, used to dye silks yellow, is a red powder, rubbed off the ripe capsules of *Rottlera tinctoria*.

83. Natural Order, *Aristolochiaceæ.*—The Birthwort Family.

Usually climbing shrubs, or herbs, with an irregular or regular perianth, valvate in bud. Ovary inferior, three- or six-celled.

TYPE—*Aristolochia indica.*

A smooth twining shrub, with alternate leaves and axillary, cymose, very irregular, brown-red and green flowers.

Organ.	No.	Cohesion.	Adhesion.
Perianth. *leaves.*	3	Gamophyllous.	Superior.
Stamens.	6	Hexandrous.	Gynandrous.
Pistil. *carpels.*	6	Syncarpous.	Inferior.

OBSERVE the one-sided expansion of the limb of the irregular perianth; in some American species, one or two of which are grown in Indian gardens, the perianth is almost large enough to form a bonnet for a child's head. In other genera of the Family the perianth is regular. It is valvate in æstivation. Observe, also, the six stamens cohering around the stigma, forming a round head at the bottom of the tube (which is often inflated) of the perianth.

84. Natural Order, *Nepenthaceæ.*—The Pitcher-plant Family.

Climbing shrubs, with alternate pitcher-bearing leaves and racemose diœcious flowers.

TYPE—Common Pitcher-plant (*Nepenthes distillatoria*).

A somewhat shrubby, climbing plant, leaves alternate with a prolonged midrib supporting a " pitcher," and terminal racemes of greenish or reddish-green diœcious flowers.

Organ.	*No.*	*Cohesion.*	*Adhesion.*
Perianth. *leaves.*	4	Polyphyllous.	Inferior.
♂ Stamens.	∞	Monadelphous.	...
♀ Pistil! *carpels.*	4	Syncarpous.	Superior.

OBSERVE the extraordinary abundance, in the stem and leaves, of spiral vessels, easily observed under the microscope, when portions of tissue are pulled asunder: the gradual development of the pitcher, which may be traced from its earliest appearance, both in seedling plants and at the tip of the prolonged midrib of the leaves. According to some botanists the "pitcher" is a hollow petiole, with its margins united in front, bearing the blade articulated to it as a "lid." Dr. Hooker's observations show that the pitcher is a glandular excavation in the end of the excurrent midrib of the leaf. The inner surface of the pitcher secretes a fluid, in which insects are frequently drowned.

This small and very anomalous Family is confined to S. E. Asia and the islands of the Indian Archipelago. In Borneo some species are found with enormous pitchers.

85. Natural Order, *Salicaceæ.*—The Willow Family.

Trees or shrubs with unisexual (diœcious) flowers. Ovary free, one-celled; ovules basal or parietal.

TYPE—Willow (*Salix tetrasperma*); or Weeping Willow of gardens (*S. babylonica*).

A small tree with alternate simple leaves and axillary catkins of achlamydeous, diœcious flowers.

Organ.	No.	Cohesion.	Adhesion.
Perianth.	o	—	—
♂ Stamens.	6-8	Hex-oct-androus.	...
♀ Pistil. carpels.	2	Syncarpous.	...

OBSERVE the arrangement of the minute flowers in deciduous spikes (*catkins*), each flower in the axil of a minute bract-scale : the one-celled ovary with two parietal placentas : the downy aril of the seeds.

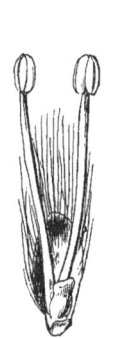 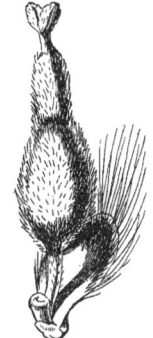 

FIG. 173. Male flower of Willow.   FIG. 174. Female flower of same.   FIG. 175. Seed of same

Cuttings of Willow, when planted in moist earth, give off adventitious roots very freely, and by this means they are easily propagated. Of the Weeping Willow of Western Asia, but one sex is in general cultivation in England. It occurs also in Indian gardens. This small Family includes but two genera, Willow and Poplar (*Populus*), both principally confined to arctic and temperate regions, and, in India and the tropics, to mountain slopes.

The twigs of species of Osier Willow are used in northern countries to make basket-work.

86. Natural Order, *Cupuliferæ.*—The Oak and Chestnut Family.

Trees with alternate leaves and monœcious flowers. Ovary inferior, three- (or more-) celled. Fruit one-celled, one-seeded.

TYPE—Spiked Oak (*Quercus spicata*).

A large timber-tree of the Himalaya and Khasia mountains, with alternate, simple, entire, coriaceous, shining, stipulate leaves, axillary, long, erect spikes of small monœcious flowers, and sessile acorn-fruits.

Organ.	*No.*	*Cohesion.*	*Adhesion.*
♀ Perianth. *leaves.*	?	Gamophyllous.	Superior.
♂ Stamens.	∓10	Decandrous.	...
♀ Pistil. *carpels.*	3	Syncarpous.	Inferior.
Seed solitary, exalbuminous.			

OBSERVE the various modifications of the involucre surrounding the acorn in different species of Oak and Chestnut (*Castanea*). In some Oaks it is closed over the fruit, splitting into valves when ripe (as in Chestnuts); in others (as is usually the case in Oaks), it forms a cup, covered with numerous imbricating scales, or prickles, or a series of rings. In other genera, not Indian, as the Hazel and Hornbeam, the involucre consists of two or three leafy bracts, which enlarge after flowering. Observe, also, the constant abortion of all the ovules contained in the ovary excepting one, as the fruit ripens, so that it is one-seeded, and, by suppression, also one-celled.

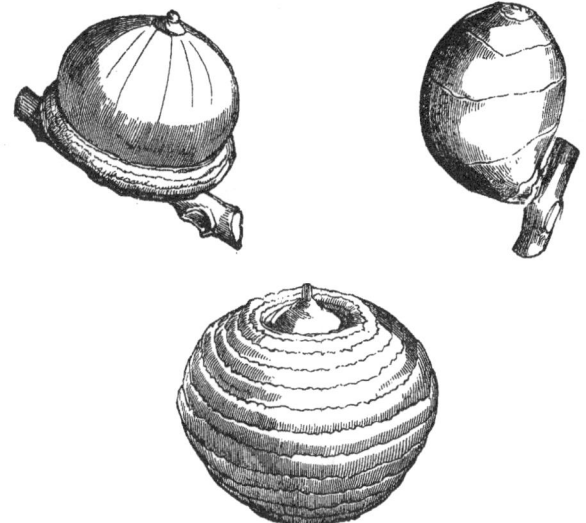

FIG. 176. Acorns of Indian species of Oak (*Quercus*).

The Family is one of very great importance in temperate countries, especially the genera Oak, Beech (*Fagus*), and Chestnut. In India, Oaks and Chestnuts are confined to the Himalaya and mountains of the Malay peninsula, where the species of Oak are very numerous, differing, generally, from the European species in their larger, often undivided, and more coriaceous leaves.

The chief value of the Family depends upon the species affording timber. Of these, by far the most important is the English Oak (*Quercus Robur*), the wood of which is invaluable wherever strength and durability are needed.

Several Oaks abound in astringent principles, and the bark of the English Oak, and acorn-cups of the Valonia (*Q. Ægilops* and allied species), are largely used in Britain by tanners. Gallic acid—used in making ink, from the intense black colour of its combination with salts of iron— and tannin are chiefly obtained from galls produced upon scrubby Oaks of Asia Minor by the puncture of insects. The gall is a diseased development of tissue induced by the puncture. Cork is the outer bark of a South European and Mediterranean Oak. It is collected at intervals of six to ten years, after the tree has attained an age of about thirty years. The bark is heated, flattened under weights, and then slowly dried. The process of barking is said not to injure the tree.

87. Natural Order, *Thymelaceæ.*—The Spurge-Laurel Family.

Shrubs with stringy bark. Ovary free, one-celled (or two-celled), with one pendulous ovule in each cell.

TYPE—Paper Daphne (*Daphne papyracea*).

A shrub (native of the Himalaya) with alternate, entire, smooth leaves, and terminal heads of regular white flowers.

Organ.	No.	Cohesion.	Adhesion.
Perianth. *leaves.*	4	Gamophyllous.	Inferior.
Stamens.	8	Octandrous.	Epiphyllous.
Pistil. *carpel.*	1 (?)	Apocarpous (?).	Superior.
Seed solitary, pendulous, exalbuminous.			

OBSERVE the remarkable tenacity of the fibre of the inner layer of the bark (*liber*), used to make a tough paper in Northern India: the stamens in two series, four above, four below.

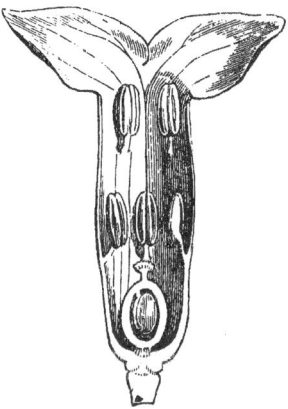 

FIG. 177. Vertical section of flower of Daphne.   FIG. 178. Vertical section of fruit of same, showing a solitary suspended seed.

The West Indian Lace-bark (*Lagetta*) belongs to this Family, nearly all the species of which are characterised by a similar tough liber. A near ally of the Daphnes is the rare Eagle-wood (*Aquilaria Agallocha*), with a fragrant resinous wood. It may be taken as a Sub-type of the Family, from which it differs in its two-celled ovary and dry dehiscent fruit.

88. Natural Order, *Santalaceæ.*—The Sandalwood Family.

Herbs, shrubs, or trees, with alternate or opposite entire leaves. Stamens opposite perianth-lobes. Ovary adherent, with ovules suspended from a free central placenta.

Type—Sandalwood (*Santalum album*).

Fig. 179 Sandalwood (*Santalum album*).

A tree with fragrant wood, opposite, entire, smooth leaves, and short, axillary and terminal panicles of small flowers, at first pale yellow, then rusty purple.

Organ.	No.	Cohesion.	Adhesion.
Perianth. *leaves*.	4	Gamophyllous.	Superior.
Stamens.	4	Tetrandrous.	Epiphyllous.
Pistil. *carpels*.	3 (?)	Syncarpous (?).	Inferior.

OBSERVE the stamens, opposite to the lobes of the perianth : small scales alternating with the stamens : the ovules attached to a free central placenta. By the latter character the Sandalwoods may be distinguished from the Mistletoe Family, to which they are nearly allied. Some of the *Santalaceæ* are parasitical.

The wood of *Santalum album* affords, on distillation, a fragrant oil much used in India as a perfume.

89. Natural Order, *Elæagnaceæ.*—The Oleaster Family.

Trees or shrubs, more or less covered with silvery scurf-scales. Base of perianth-tube persistent over a free, one-celled ovary. Ovule erect.

TYPE—*Elæagnus* (any Indian species).

Shrubs or trees, more or less covered with minute shining scales, with alternate, entire leaves, and small, axillary, fascicled, regular flowers.

Organ.	No.	Cohesion.	Adhesion.
Perianth. *leaves.*	4	Gamosepalous.	Inferior.
Stamens.	4	Tetrandrous.	Epiphyllous.
Pistil. *carpel.*	1 (?)	Apocarpous (?).	Superior.
Seed solitary, erect.			

OBSERVE the scurf-like peltate scales, which more or less abundantly cover all the young parts and the under surface of the leaves ; scraped off with a knife, they are interesting objects under the microscope : the base of the perianth-tube, which persists as in *Nyctagineæ*, closely investing the

U

ovary as a false pericarp.   In some species the fruit is succulent; that of *E. orientalis* is used in dessert under the name of Trebizonde Dates.

The Family is a very small one, and principally confined to the Old World.

90. Natural Order, *Myristicaceæ.*—The Nutmeg Family.

Trees or shrubs, with alternate entire leaves, and inconspicuous diœcious flowers.   Fruit one-celled, one-seeded. Albumen ruminated.

TYPE—The Nutmeg-tree (*Myristica fragrans*)

An introduced tree, thirty to forty feet high, with alternate entire leaves, small, yellowish, supra-axillary, diœcious flowers, and one-seeded, drupaceous (dehiscent) fruits.

Organ.	No.	Cohesion.	Adhesion.
Perianth. *leaves.*	3	Gamophyllous.	Inferior.
♂ Stamens.	9-12	Monadelphous.	...
♀ Pistil. *carpel.*	1	Apocarpous.	Superior.
Seed solitary, arillate, with ruminated albumen.			

OBSERVE the acrid juice of the bark, staining red : the scarlet aril, forming a coarse network over the seed, laid bare, by the fleshy pericarp dehiscing, when ripe, in two valves : the albumen, interrupted by folded plates of a different tissue ; such albumen is said to be ruminated (*see* Custard-Apple Family, p. 151) : the spreading (divaricate) cotyledons of the minute embryo.

A small tropical Family, including but few species of importance besides that employed as Type, the seed of which is the Nutmeg of commerce, and the aril Mace. The Nutmeg-tree is said to be indigenous in the Moluccas, though now widely dispersed in the Indian Archipelago. Aromatic fruits, more or less resembling the Nutmeg, are afforded by several other species of the genus *Myristica*.

91. Natural Order, *Lauraceæ.*—The Laurel Family.

Trees or shrubs, with entire, usually evergreen leaves (*Cassytha* is leafless and parasitical). Anthers opening by valves. Seed solitary, exalbuminous.

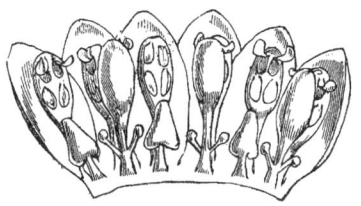

FIG. 180. Perianth of *Cinnamomum* laid open, showing the stamens and staminodes.

TYPE—Cinnamon Shrub (*Cinnamomum zeylanicum*).

A small tree or shrub, with subopposite, entire, smooth, coriaceous, three-nerved leaves, and loose terminal and axillary panicles of small, greenish-white flowers.

Organ.	No.	Cohesion.	Adhesion.
Perianth. *leaves.*	6	Gamophyllous.	Inferior.
Stamens.	9	Enneandrous.	...
Pistil. *carpel.*	1	Apocarpous.	Superior.
Seed solitary, exalbuminous.			

U 2

OBSERVE the three staminodia, alternating with the innermost stamens : the dehiscence of the anthers by four cells,

each opening by a recurved valve, as in the Barberries : three of the nine perfect stamens, with lateral glands attached to their filaments, have the valves· of their anthers opening inwards.

The genus *Cassytha* is a singular Sub-type, bearing the same relation to the rest of the Family that *Cuscuta* bears to the Convolvulus Family. The species are leafless twining parasites. The flowers differ but slightly from those of the true Laurels.

In some Indian genera (*Litsaea,* *Tetranthera,* &c.) the flowers are dioecious, and in clusters or umbels. The Laurel Family is chiefly confined to tropical countries ; several species, however, extend as far northward as Japan in Eastern Asia, and one, the Victor's Laurel (*Laurus nobilis*), reaches the South of Europe.

FIG. 181. An Indian species of *Cassytha,* somewhat reduced; with a fragment of the plant upon which it preys.

Many species are aromatic ; none more so than our Type, the bark of which is the Cinnamon of commerce.

Allied species of *Cinnamomum* afford Cassia. Camphor is obtained by distillation from the wood of a Chinese species of *Cinnamomum* (*C. Camphora*).

92. Natural Order, *Piperaceæ.*—The Pepper Family.

Jointed herbs or shrubs with alternate or opposite simple

leaves.   Flowers in dense spikes or racemes.   Perianth o.
Ovary one-celled, one-ovuled.

TYPE—The Betle (*Chavica Betle*).

A smooth, woody, climbing or creeping, jointed plant
giving off numerous adventitious roots, with alternate,
cordate, seven- or nine-nerved leaves, and leaf-opposed,
drooping catkins of minute, diœcious flowers.

OBSERVE the flowers, borne in the axils of minute, peltate,
shortly-pedicelled bracts : the minute embryo and double
albumen.

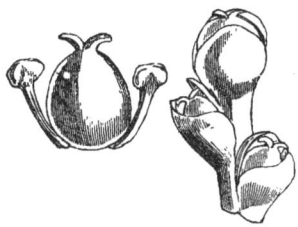

·FIG. 182. Portion of spike and detached hermaphrodite flower of Pepper (*Piper*).

The Peppers are almost wholly a tropical Family,
abounding in the hottest regions of South America, India,
and the Indian islands.   They generally agree in habit
with our Type, though some of them are low, prostrate
herbs.

Many species are aromatic, or pungent and biting.   The
Type-species is universally cultivated in India as a masti-
catory, being chewed with lime and the nut of the Areca
Palm.   The root of *Piper methysticum* is similarly employed
in the Pacific islands under the name of Kava.

Black pepper is the unripe, dried berries of *P. nigrum ;*
white pepper the same berries allowed to ripen, with the

pulpy coat removed. Long pepper is the dried flower-spikes of *Chavica Roxburghii.*

The Family affords some valuable medicines, and several species are employed as such by the natives of India.

93. Natural Order, *Coniferæ.*—The Pine Family.

Branching trees with simple, usually evergreen leaves. Ovules naked (fertilised by direct contact of the pollen).

TYPE—The Cheer Pine (*Pinus longifolia* [Cheersullah, Sarut]).

A tree of the Himalaya, with long, acicular, evergreen, ternate leaves, naked, monœcious flowers in catkins, and a multiple fruit (cone).

[No species of Pine is native in the Peninsula.]

From the extreme simplicity of the flowers of Coniferæ, the usual schedule is not suited to exhibit their structure in a tabular form.

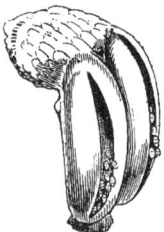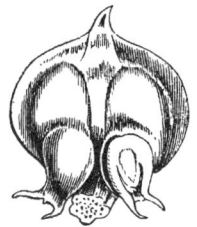

FIG. 183. Staminal scale of Pine.   FIG. 184. Ovule-bearing scale of same.

The male flowers of Pine (*Pinus*) are arranged in short catkins, consisting of minute, imbricating scales, each scale bearing two anther-lobes upon its under surface.

The female flowers, also, are in small, dense, cone-like catkins, consisting of small scales, each scale bearing upon the base of its upper side a pair of inverted ovules. As the scales are closely imbricated, the ovules are concealed ; but they may be easily found by breaking the flowering cone across the middle, when some of them are sure to be exposed.

Some botanists are of opinion that the scales to which the ovules are attached are open carpellary leaves. Their true nature is not satisfactorily settled. In any case, the ovules are naked, so that the pollen-grains fall directly upon the ovules. Hence the term *gymnospermous* applied to the Family, in contradistinction to *angiospermous* applied to all other flowering plants in which the ovules are fertilised through the medium of the stigma of a carpellary leaf. In the fruit the ovule-bearing scales are much enlarged, and hard and woody in texture, each scale bearing upon its upper surface a pair of winged seeds.

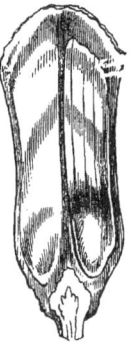

FIG. 185.
Scale of cone of Pine, bearing two winged seeds.

The scales, both of flower and fruit, are arranged upon common axes in the form of a cone ; hence the name Coniferæ applied to the Pine Family.

In Cypress (*Cupressus sempervirens*, an exotic species), Yew (*Taxus baccata*), and Juniper (*Juniperus communis*), this Type is slightly departed from, though all agree in the naked ovules of the female flowers.

In Cypress, the scales of the male catkins bear four anther-cells, and the ovules are numerous and erect in the axils of a small number of scales arranged in a head. These scales become woody and peltate, constituting a modification of the cone called a *galbulus*.

In Yew, the male flowers consist of peltate scales, bear-

ing about six (three to eight) anther-cells ; the female flowers
of solitary ovules, around each of which a succulent, pink-
coloured disk developes as they mature, enclosing and over-
topping the fruit.

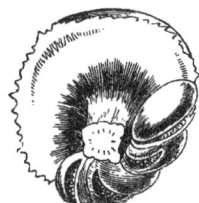

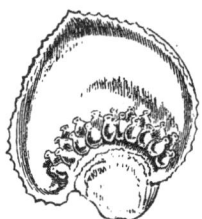

FIG. 186. Scale of male inflorescence
of Cypress.

FIG. 187. Scale of female inflorescence
of same.

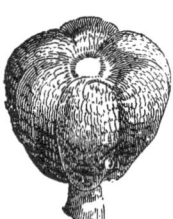

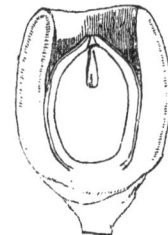

FIG. 188. Stamen
of Yew.

FIG. 189. Female flower
of same.

FIG. 190. Vertical section
of fruit of same.

In Juniper, the anther-scales are four-celled, and the
ovules three in number, one at the base of each of three
connate scales, which form a succulent galbulus when the
seeds are ripe.

Besides the peculiarity in the structure of the female flower
of Coniferæ, the Order is characterised by the absence of
vessels in the wood, which consists of tapering wood-cells,
marked on the sides, towards the medullary rays, with

circular disks, which answer to the margins of minute, lenticular, intercellular cavities occurring between the " pits " of adjacent cells. In the Pines, and allied species, the cotyledons are usually numerous, varying from three to

FIG. 191. Female inflorescence of Juniper.

FIG. 192. Seed of same, bearing a few resin-receptacles.

eighteen; hence the term *polycotyledonous* applied to them. As in other respects the structure of the Coniferæ approaches that of Dicotyledons, they are usually classed along with them as an anomalous Family.

OBSERVE, also, the different forms and the arrangement of the leaves in Coniferæ. In Scotch Fir, for example, there are two forms of leaf, viz. small, brown, scaly leaves on the main branches, the internodes of which lengthen out, and in the axil of each of these scaly leaves a single pair of long acicular leaves, sheathed at the base by scale-leaves. The long acicular leaves are borne upon axillary arrested branches.

In some other species of Pine the arrested branches bear the acicular leaves in fascicles of threes or fives.

In Larch (*Larix*) and Cedar (*Cedrus*) the acicular leaves are numerous, in dense fascicles. The former species is well adapted to show the true nature of these fascicles of leaves, some of which lengthen out into branches during the summer. Indeed, the fruit-cones occasionally lengthen out in this way into leaf-bearing branches, illustrating the homology subsisting between the bract-scales of the cone and the scale-leaves of the branch.

The leaves of many species of Pine persist several years. The Larch is deciduous.

The Pine Family acquires high importance from the number of species which it includes affording valuable timber, and also from the resinous products obtained from several of them.

Differing very much in habit from the Type, and generally treated as a distinct Family, is the group *Gnetaceæ*, represented in India by species of *Gnetum* and *Ephedra*, the latter confined to the Himalaya.

The Gnetums are climbing shrubs with jointed stems, opposite, shining, entire leaves, and axillary spikes of verticillate, monœcious flowers. The male flowers each consist of a single stamen ; the female, of a naked ovule, terminating above in a long, tubular prolongation of its coat, resembling a style, and enclosed in an undivided perianth. The fruit is an oblong drupe ; in at least one species edible.

The Ephedras are leafless, much-branched shrubs, growing in desert regions of the temperate zones. Associated with *Gnetum* and *Ephedra* is the genus *Welwitschia*, a most extraordinary and very anomalous dwarf tree of South Africa, attaining a great age, with a table-like trunk seldom raised more than six or twelve inches above the sand in which it grows, and a single pair of leaves, persisting through the lifetime of the tree, and believed to be the cotyledonary leaves.

94. Natural Order, *Cycadaceæ.*—The Cycas Family.

Unbranched, palm-like trees with a terminal crown of pinnate leaves.

TYPE—*Cycas revoluta.*

FIG. 193. *Cycas revoluta.*

An unbranched, palm-like tree, sometimes ten to twelve feet high, common in gardens, bearing a crown of pinnate leaves with numerous, rigid, revolute-margined leaflets, and dioecious flowers in large terminal cones.

Differing principally from the Pine Family in the simple stem, marked with the scars of fallen leaves, the structure of the wood, and the pinnate leaves.

The scales of the male cones, which correspond to single anthers of ordinary flowers, bear upon their under surface very numerous, scattered, and clustered cells containing pollen-grains. The ovules are borne upon the margins of the scales of the female cones. From the seeds of an Indian species commonly planted in Malabar and Ceylon a

useful flour is obtained, which is used by the poorer natives.
From the pith of some other species a coarse sago is collected. Fossil remains indicate that Cycadeous plants were
very abundant in Europe at the time of the deposition of the
chalk, and during the early tertiary period. They are now
principally confined to Mexico, South Africa, and Australia.
Our Type-species of the Family is a northern outlier at the
present period, native in China and Japan.

MONOCOTYLEDONS.

SPADICIFLORÆ.

95. Natural Order, *Palmaceæ.*—The Palm Family.

Stem woody. Perianth six-leaved. Leaves pinnately or
palmately divided.

TYPE—Cocoa-nut Palm (*Cocos nucifera*).

A tall, unbranched tree, with a terminal plume of large,
pinnate leaves, much-branched spadices of small, monœcious
flowers from the axils of the outer leaves of the crown, and
large fruits (Cocoa-nuts), with a fibrous epicarp.

Organ.	*No.*	*Cohesion.*	*Adhesion.*
Perianth. *leaves.*	6	Polyphyllous.	Inferior.
♂ Stamens.	6	Hexandrous.	...
♀ Pistil. *carpels.*	3	Syncarpous.	Superior.
Seed large, solitary, albuminous.			

OBSERVE the germination of a Palm, the sheathing portion of the cotyledon often penetrating, at the expense of

the large store of albumen, to a considerable depth in the
soil before the development of the plumule : the woody stem,
varying in height and form in different species, sometimes
prostrate, forming a rhizome, or cable-like, but usually erect
and nearly cylindrical, bearing the persistent bases of fallen
leaves, or exhibiting the ring-like scars which they leave. It is
rarely branched, as in the Egyptian Doum Palm (*Hyphaene*),
an erect-growing species. The wood is often extremely hard
outside, and very soft within. It is well adapted to illustrate
the independence of the vascular cords, characteristic of
Monocotyledons (see page 115). Observe, also, the variety
in structure of the fruit ; the three carpels of which it is
normally composed are usually coherent, but sometimes,
as in *Chamærops, Rhapis,* and *Phœnix* (the Date), the fruit
is apocarpous. In the more important Indian genera it is
syncarpous, and one-celled from the suppression of two
cells, as in *Cocos* and the Betel (*Areca*),—or three-celled, as
in *Arenga, Caryota, Calamus, Borassus,* and others. The
structure of the pericarp is particularly variable. In the
Cocoa-nut (*Cocos*) the epicarp is fibrous, affording the coir-
fibre of commerce ; the endocarp, a hard shell. In the
Date (*Phœnix*) the pericarp is fleshy and sweet. In the
Rattans or Rotangs (*Calamus*), and Sago Palm (*Sagus*), the
fruit is covered with numerous, hard, imbricating scales.
In the Palmyra Palm (*Borassus*) it is a huge drupe, with
three large, fibrous, one-seeded pyrenes.

The useful products of this princely Family are so infinitely numerous that a few only of the more important,
afforded by Indian species, can be referred to here.

The wood of several species is used in building and other
constructions. The hard outer portion of the trunk is
exported to Europe for walking-sticks and umbrellas.
Canes or Rattans, the flexible stems of species of *Calamus*,

are also largely exported from the Malay peninsula, for use in seating chairs, &c.   The pith-like tissue of the inside of the trunk of *Sagus*, growing in the Archipelago, affords abundant farinaceous matter, from which sago is prepared for home use and exportation.   The saccharine juice of species of *Phœnix, Borassus, Caryota,* and other genera, is collected and fermented as palm-wine, distilled for arrack, or boiled down for sugar.

Of the leaves numerous applications are made.   Strips are worked up into baskets, and punkahs are made of the large fan-leaves of the Palmyra (*Borassus*).   Books are made of strips of the leaves of the Palmyra and Talipot (*Corypha*) Palms.

The Cocoa-nut, collected as food and for the sake of its excellent oil, and the astringent seeds, with ruminated albumen, of the Betel (*Areca Catechu*), universally chewed in tropical Asia, are the most important Indian fruit-products of the Family.   The resin called Dragon's-blood is principally obtained from a species of *Calamus.*

Palms are almost exclusively tropical, abounding in the hot and humid parts of Asia and South America.   The species are proportionally less numerous in Africa, though on the western coast of that continent grows one of the most useful members of the group,—the Oil Palm (*Elaïs*). The oil obtained from the fruit is largely consumed in England in the manufacture of soap and candles.   The species are generally restricted in the area over which they extend.   The Cocoa-nut is one of the most widely dispersed, occurring on the shores of most tropical countries.   A few Palms reach as far north as China, Japan, and the United States, while a single species is native of Southern Europe,—the Dwarf Fan Palm (*Chamærops humilis*).

96. Natural Order, *Pandanaceæ.*—The Screw-Pine Family.

Stem woody or herbaceous.   Leaves linear (except *Nipa*). Flowers sessile, in heads or spikes.   Perianth o (except in *Nipa*, ♂).

TYPE—*Pandanus odoratissimus.*

FIG. 194. Screw-Pine (*Pandanus odoratissimus*), showing aerial roots from the lower part of the trunk.

A forking or unbranched tree, everywhere planted, giving off buttress-like adventitious roots, with a terminal crown of long, prickly leaves arranged in three spiral rows, and dioecious flowers: the males delightfully fragrant, in long, pendulous, leafy, panicled spikes; the female in a terminal one.

The flowers are achlamydeous, the stamens being crowded upon the spadices, and often cohering in bundles; the carpels one-celled and one-ovuled, densely packed on the female cone. The multiple fruit consists of a number of closely-packed fibrous one-seeded drupes.

The Screw-Pines derive their appellation not from any resemblance to the Pine Family, but rather from the similarity of their foliage to that of the Pine-apples (*Bromeliaceæ*), and especially to that of the Pine-apple itself (*Bromelia Ananas*), a tropical American plant much cultivated in hot climates.

OBSERVE the tendency to form adventitious roots from the lower part of the trunk : the forked branching of the trunk, unusual amongst arborescent Monocotyledons : the trifarious (three-rowed) arrangement of the leaves, with their prickly margins and keel.

The leaves make a good thatch, and are used for matting. The fibrous roots serve as cordage, and also, when cut up, as corks.

A near ally of the Screw-Pines is the dwarf, palm-like *Nipa fruticans*, abundant in the Sunderbunds. The leaves are pinnate, and the small flowers are provided with six-leaved perianths.

97. Natural Order, *Typhaceæ.*—The Bullrush Family.

Marsh herbs, with linear leaves, and spicate or capitate, monœcious flowers. Perianth o. Fruit a dry one-seeded nut.

TYPE—*Typha elephantina.*

A perennial herb, growing on the borders of tanks and lakes, with radical ensiform leaves, and a tall scape terminating in a cylindrical inflorescence, of which the lower portion consists of female, the upper, separated by a short interval, of male flowers, very densely packed.

The flowers are achlamydeous; the males consisting of two to four stamens, the female of a pedicellate pistil, with a one-celled ovary, surrounded by a whorl of hair-like filaments, representing a perianth.

The leaves are used for matting, and are said to be tied up in bundles to serve as swimming floats. The pollen is collected, and made up into cakes, and eaten as bread, in Western India.

98. Natural Order, *Aroideæ.*—The Arum Family.

Stem herbaceous, or woody, or wanting, with leaves usually net-veined. Flowers monœcious (sometimes diœcious or hermaphrodite), sessile on a spadix. Perianth usually o, or of minute scales.

TYPE—Kuchoo or Kachalu (*Colocasia antiquorum*).

A stemless, perennial herb, extensively cultivated, with large, radical, peltate, arrow-head leaves and monœcious achlamydeous flowers arranged upon a fleshy spike (*spadix*), enclosed in a yellowish sheathing bract (*spathe*).

The lower portion of the spadix is occupied by numerous female flowers, each consisting of pistil only, with one-celled ovary and several ovules upon two or three placentas. Adjoining and above the female flowers are some abortive pistils, then a number of closely-packed male flowers, each reduced to a single two-celled anther, opening by minute pores at the top. The anthers cohere, side by side, in masses. The spadix is prolonged beyond the crowded stamens into an acute "appendix," which takes no part in

x

reproduction, and which is absent in the allied, ornamental
"Lily of the Nile" (*Richardia æthiopica*) of gardens.

The entire spadix and spathe are liable to be mistaken
by beginners for a single flower; but a comparison of our
Type-species with other genera removes all doubt, and
proves the spadix to be an inflorescence, bearing innume-

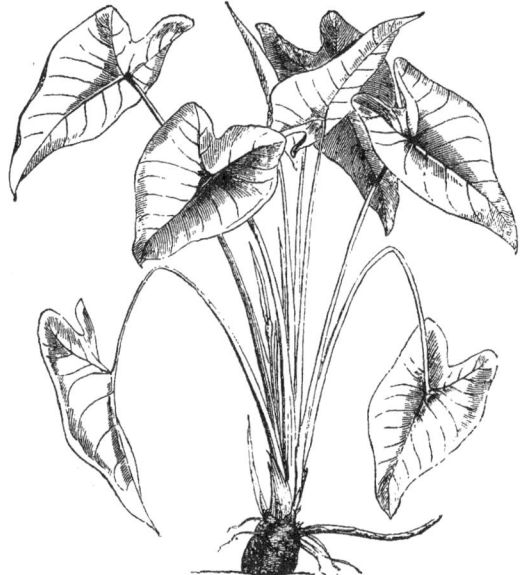

FIG. 195. *Colocasia antiquorum*, much reduced.

rable flowers. In the *Richardia* of South Africa (not of
the Nile, as its familiar name would imply,) the pistils are
each surrounded by three abortive stamens (*staminodia*),
and in the common Indian climbing *Pothos* the flowers
are hermaphrodite, each with a six-leaved perianth.

The Type-species, and several allies of the same and of different genera, are very valuable food producing plants. widely cultivated in the tropics.

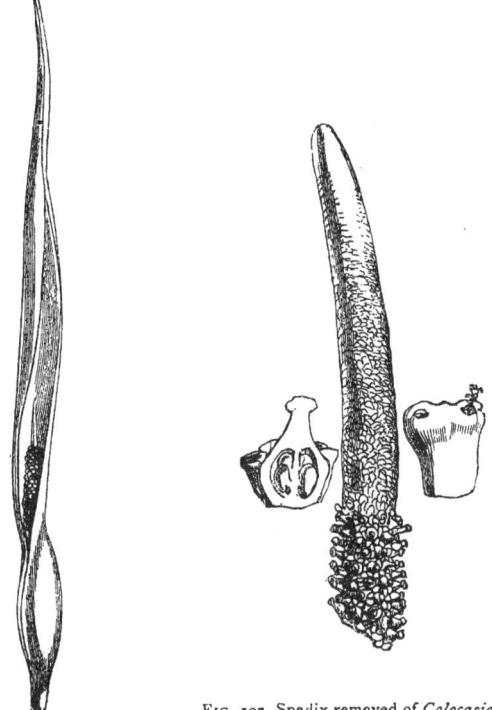

FIG. 196. Spathe and spadix of *Colocasia antiquorum*.

FIG. 197. Spadix removed of *Colocasia*. A detached stamen to the right; a pistil and surrounding scales to the left.

In India the *Colocasia* is propagated by offsets from its tubers, which grow to a large size, and contain abundance of excellent farinaceous matter, from which the acrid juice, characteristic of the Family, is driven off by the process of

X 2

cooking. It is a variable plant, as are most largely cultivated species, and some of its varieties have been separated
as distinct species. The aromatic *Acorus Calamus,* with
sword-shaped leaves, common in Indian gardens, is a very
widely dispersed member of the Family; occurring in England, through Europe and temperate Asia, and also in North
America. The fragrant rhizome is chewed, made up as a
confection, or used medicinally.

The leaves of *Colocasia* and some of its allies, growing
in swampy places, distil water from a minute pore at the
tip, to which point free canals, in the substance of the leaf,
converge. This process supplements the transpiration from
the surface of the leaf, and is most abundant when transpiration is checked by the moisture of the atmosphere.

99. Natural Order, *Pistiaceæ.*—The Duckweed Family.

Floating herbs, in *Lemna* consisting of minute, leaf-like
fronds.

TYPE—*Pistia Stratiotes.*

A floating herb, with tufted obcordate-
cuneate radical leaves, numerous fibrous
roots hanging in the water, and minute
spathes with adherent spadices rising
from the middle of the tuft.

The annexed cut will explain the inflorescence of *Pistia* better than a verbal
description. The spadix is adherent to
the spathe, and terminates in a small
head of five adherent four-celled anthers.
Immediately below the head of the anthers
is the stigma. The ovary is one-celled and
adherent, containing several ovules.

FIG. 198. Spathe of
Pistia, in longitudinal
section.

This little lettuce-like plant is gigantic

compared with the minute representatives of the Family in Europe, belonging to the genus Duckweed (*Lemna*). Their fronds are everywhere common, floating in stagnant water. Some species of *Lemna* occur in India.

100. Natural Order, *Taccaceæ.*—The Tacca Family.

Herbs, with radical leaves and scapes, bearing umbellate flowers.

TYPE—*Tacca pinnatifida.*

A perennial herb, cultivated in Southern India, with large, tripartite, radical leaves with pinnatifid lobes, and long scapes bearing an involucrate umbel of greenish flowers.

Organ.	No.	Cohesion.	Adhesion.
Perianth. leaves.	6	Gamophyllous.	Superior.
Stamens.	6	Hexandrous.	Epiphyllous.
Pistil. carpels.	3	Syncarpous.	Inferior.

This species is cultivated all through Polynesia for the sake of its mealy tubers. The Family is a very small one, and confined to the tropics of the Old World.

PETALOIDEÆ.

101. Natural Order, *Dioscoreaceæ.*—The Yam Family.

Usually twining herbs, with net-veined, simple or digitate leaves. Flowers unisexual. Perianth six-lobed. Ovary inferior, three-celled.

TYPE—*Dioscorea sativa* (or any other species of Yam).

A twining herb, with alternate, more or less cordate leaves, axillary spikes of very small dioecious flowers; the males in slender panicles, the females in simple spikes ; and three-lobed, capsular fruits.

Organ.	*No.*	*Cohesion.*	*Adhesion.*
Perianth. *leaves.*	6	Gamophyllous.	Superior
♂ Stamens.	6	Hexandrous.	Epiphyllous.
♀ Pistil. *carpels.*	3	Syncarpous.	Inferior

A small Family, widely spread through hot countries ; one species with berried fruits representing it in Britain. The species much resemble the Sarsaparilla Tribe of the Lily Family in habit, with which they agree in having net-veined leaves. The inferior ovary, however, at once distinguishes them. The species are generally acrid, but, in those affording the large tuberous roots called Yams, this acridity, when present, is removed by cooking. Several variable species are cultivated for Yams in nearly all tropical countries. Their culture is believed to have spread from South-Eastern Asia and the Archipelago.

OBSERVE the minute green bulbels often borne in the axils of the leaves of the Type-species.

102. Natural Order, *Liliaceæ.*—The Lily Family.

Herbs (in *Dracæna* shrubs or trees), with a six-leaved petaloid perianth. Ovary superior, three-celled.

TYPE—*Dracæna ferrea.*

Organ.	*No.*	*Cohesion.*	*Adhesion.*
Perianth. *leaves.*	6	Gamophyllous.	Inferior.
Stamens.	6	Hexandrous.	Epiphyllous.
Pistil. *carpels.*	3	Syncarpous.	Superior.

An erect, shrubby or arborescent (Chinese) plant, common in Indian gardens, with terminal crowns of red-brown leaves, and large, terminal panicles of small, white or purplish, racemose flowers.

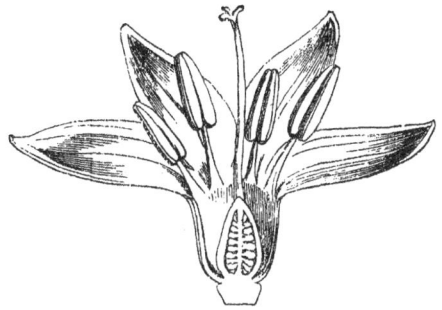

FIG. 199. Flower of *Dracæna*, in longitudinal section.

A large Family, including several marked Sub-types, differing from each other in habit rather than in the structure of their flowers. They do not form a conspicuous feature in Indian vegetation; several, however, are greatly prized in

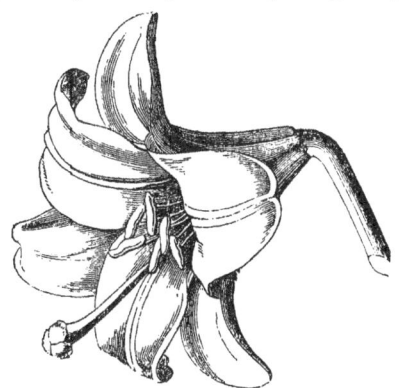

FIG. 200. Flower of Lily (*Lilium*).

gardens for the beauty of their flowers. Amongst the latter few are more showy than the *Gloriosa* (*Methonica*) *superba* of Indian forests, with its climbing, herbaceous stem, aided by tendrils terminating the leaf-blades.

The African genus *Aloe* (not the so-called American Aloe (*Agave*) belonging to the Amaryllis Family) and the Adam's Needle (*Yucca*) belong to the Lily Family, as do the cultivated herbs Garlic, Onion (*Allium*), and Asparagus.

Bowstring Hemp is a tenacious fibre obtained from the leaves of an Indian species of *Sanseviera*. *Phormium tenax* of New Zealand yields a similar most valuable fibre.

Peliosanthes Teta, — a stemless perennial, with plaited radical leaves, and small, green, racemose flowers,—represents an Indian Sub-type in which the ovary is partially inferior. The Sarsaparillas (*Smilax*) form a Tribe, sometimes regarded as a distinct Family, differing from the true Lilies in having net-veined leaves, climbing stems, and small diœcious flowers. They resemble the Yams (*Dioscorea*) in appearance. The species are widely spread, and several occur in India.

103. Natural Order, *Juncaceæ.*—The Rush Family.

Herbs. Perianth-leaves scarious. Stamens six. Ovary superior.

TYPE—*Juncus bufonius.*

A small, tufted, much branched, annual weed, with inconspicuous, solitary, or fascicled flowers scattered along the stems in the axils of slender leafy bracts.

This small Family scarcely differs from the preceding, excepting in the dry texture of the small six-leaved perianth, and in the very minute embryo. The species employed as Type is a common, very widely spread weed, in places liable to inundation and near water, especially in temperate countries, to which, indeed, the Rush Family is mainly confined. The Type differs much in habit from other common species of the same genus, in most of which the inflorescence forms a small panicle, either terminating the slender cylindrical stem, or apparently given off from the side of it. The leaves of Rushes are often transversely divided by plates of pith, so that they seem to be jointed.

104. Natural Order, *Commelynaceæ.*—The Spiderwort Family.

Herbs. Three outer leaves of perianth herbaceous, inner petaloid. Ovary superior, usually three-, sometimes two-celled.

TYPE—*Commelyna benghalensis.*

A branched, creeping, more or less hairy, perennial herb, with sheathing leaves and hooded bracts, enclosing one male and two or three hermaphrodite, bright blue flowers.

Organ.	No.	Cohesion.	Adhesion.
Perianth. *leaves.*	6	Polyphyllous.	Inferior.
Stamens.	6	Hexandrous.	Hypogynous.
Pistil. *carpels.*	3	Syncarpous.	Superior.

In this Family, as in the Water Plantains, the perianth consists of outer sepaloid and inner petaloid segments, as is usual in Dicotyledons. In most Monocotyledons with conspicuous flowers all the leaves of the perianth are petaloid.

OBSERVE the filaments in some genera (*Cyanotis,* &c.), bearded (*stupose*), with moniliform hairs, in the cells of which the movement of currents of viscid protoplasm may be observed under a high magnifying power : the minute embryo embedded in a cavity at one side of the albumen.

Very few species of this widely dispersed (but not large) tropical or sub-tropical Family are turned to any account.

105. Natural Order, *Eriocauloneæ.*—The Pipewort Family.

Aquatic or marsh herbs. Flowers minute, unisexual, in terminal heads.

Type—*Eriocaulon sexangulare* (or *E. quinqueangulare*).

A small aquatic herb of rice-fields and wet places, with narrow, grass-like leaves, and small, involucrate heads of minute monœcious flowers.

Organ.	No.	Cohesion.	Adhesion.
Perianth. *leaves.*	6	Gamophyllous. (two series.)	Inferior.
♂ Stamens.	6	Hexandrous.	Epiphyllous.
♀ Pistil. *carpels.*	3	Syncarpous.	Superior.

Eriocaulon is the only large natural genus of the Family. It is very widely spread in both hemispheres, though particularly abundant in South America. The excessively minute flowers are arranged, like the florets of Compositæ, in terminal heads, borne by slender scapes.

Very few species are made use of by man.

106. Natural Order, *Pontederiaceæ.*—The Pontederia Family.

Aquatic herbs. Flowers petaloid, racemose, from the sheath of the last, or only leaf, of the scape.

TYPE—*Monochoria vaginalis.*

An aquatic herb, common in rice-fields and ditches, with radical, petiolate, cordate leaves, and racemes, apparently springing from the side of a petiole, of several rather large bright blue flowers.

Organ.	No.	Cohesion.	Adhesion.
Perianth. *leaves.*	6	Polyphyllous.	Inferior.
Stamens.	6	Hexandrous.	Epiphyllous.
Pistil. *carpels.*	3	Syncarpous.	Superior.

OBSERVE the raceme, borne upon a one-leaved scape ; as the petiole is directly continuous with the scape, while the raceme is lateral, the latter appears to spring from the side of a petiole.

This species is employed for various purposes in Indian medicine.

The Family is a very small one, chiefly confined to the stagnant waters of hot countries.

107. Natural Order, *Orchidaceæ.*—The Orchid Family.

Epiphytal or terrestrial herbs, with irregular flowers. Stamen 1 (except *Cypripdieæ*), anther gynandrous. Ovary inferior.

TYPE—*Dendrobium nobile.*

Organ.	No.	Cohesion.	Adhesion.
Perianth. *leaves.*	6	Gamophyllous.	Superior.
Stamen.	1	Monandrous.	Gynandrous.
Pistil. *carpels.*	3	Syncarpous.	Inferior.

An epiphytal herb, cultivated in gardens, with pendulous branches, emarginate leaves, and beautiful pale-sulphur or white and rose-coloured irregular flowers, with a purple eye.

FIG. 201. Flower of *Dendrobium nobile*, natural size.

The three outer leaves of the perianth are often called sepals, and the three inner leaves petals, in this Family. Of the three inner leaves, two are lateral and equal, and one (usually the lower one from the twisting of the ovary) different in form, often much larger than the sepals and lateral petals, sometimes lobed and jointed, sometimes very small. This odd petal is called the *labellum*. It is sometimes provided with a *spur* at its base containing nectar, much sought after by insects; the visits of which are in many cases absolutely necessary, in order that the flowers may be fertilised and good seed produced.

COMPARE the structure of the anther and pollen of *Dendrobium* with that of the Sub-types—

1. *Vanda Roxburghii,* an epiphyte, frequent on the Mango and other trees, with distichous, recurved leaves, and loose axillary racemes of chequered yellowish and purple flowers.

2. *Platanthera Susannæ,* an erect "terrestrial" herb with sheathing leaves, and a few very large white irregular flowers, with the lateral lobes of the spurred lip deeply fringed.

In each of these Sub-types, as in the Type-species, there is but a single stamen, adherent to the stigma, or to a continuation of the pistil immediately above the stigma, called the *column.*

The anther is sessile and two-celled, each cell containing the pollen-grains cohering together into a waxy "pollen-mass" called a *pollinium.*

In *Platanthera* the two cells of the anther are erect and nearly parallel, diverging a little below. Each cell contains a club-shaped pollinium, connected below with a slender stalk called the *caudicle.* The caudicles terminate in minute disks or knobs, which nestle in a projection of the column immediately over the opening into the

Fig. 202. *Platanthera Susannæ,* much reduced.

long spur of the labellum. The pollinia of *Platanthera* are quite separate from each other, and with their caudicles and disks may be independently removed from the anther-cells.

In *Vanda* the two pollinia are connected to a single " pedicel " (as Mr. Darwin calls it), which pedicel is attached to a viscid disk at its lower end.

In *Dendrobium* the small two-celled anther is terminal, the anther-case forming a minute cap at the top of the column. It encloses four pollinia, connected in pairs, one pair in each cell of the anther. They are not provided with either a caudicle or viscid disk.

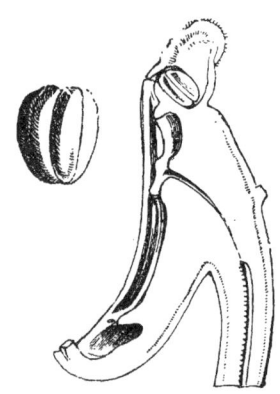

FIG. 203. Pollinia, caudicle, and gland of *Vanda.*   FIG. 204. Longitudinal section of column of *Dendrobium.* To the left the pollen detached.

From observations which have been recently published by Mr. Charles Darwin,* it is shown that the peculiar modifications of the pollen characteristic of Orchids stand in relation to the part fulfilled by insects in securing their fertilisation. As his observations have not been generally extended to Indian species, I shall briefly describe the structure and mode of fertilisation in a common British Orchis, presenting, in the structure and relative position of

* " On the Fertilisation of Orchids." Murray.

its anther, much similarity to that of the beautiful Indian *Platanthera* referred to above.

In the common Spotted Orchis of English meadows the anther is two-celled, the cells being parallel, and each containing a distinct pollinium, with caudicle and disk, as in *Platanthera.* Both of the disks rest in a small, round, knob-like projection (the *rostellum*) at the base of the column and immediately over the viscid stigma and spur of the labellum.

Take the very fine stem of a grass or a finely-pointed pencil, and thrust it gently into the spur of a newly-expanded flower, which has not lost its pollen, just as an insect would insert its proboscis when in search of nectar. It will be found that the pencil does not fail to push against the projecting rostellum, so that the pouch-like membrane of the latter is pressed down, and the pencil

FIG. 205. Pollinium of Orchis.

FIG. 206. Flower of Spotted Orchis.

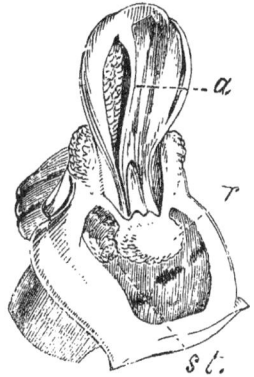

FIG. 207. Column of Spotted Orchis *a* anther ; *r* rostellum ; *st* stigma.

comes in contact with the under viscid surface of one or of both of the little glands of the two pollen-masses.   On withdrawing the pencil, the pollinia are found adhering firmly to it, for the viscid substance which bathes the glands sets hard in a few seconds when exposed.   If the pollinia be carefully watched *immediately* after they are withdrawn from the anther, they may be observed to become inclined forwards to such an extent, that if, after the lapse of a minute or two, the pencil be thrust into the nectary of a second flower, the pollinia which adhere to the pencil will strike against the viscid stigmatic surface of the flower, and at least a portion of the pollen-grains will adhere to it and fertilise the ovules of the flower.   The viscidity of the stigma is sufficient to overcome the strength of the delicate threads which bind the grains of pollen together.

From the peculiar relative arrangement of the pollen-masses and stigma in Orchids, Mr. Darwin has shown that the flowers can be but very exceptionally self-fertilised. Almost invariably insect aid is required to transport the pollen from flower to flower; hence the importance of the contrivances indicated above (to which Mr. Darwin has recently directed attention), to insure the proper fulfilment of the important function assigned to unconscious agents.

It is extremely desirable that similar observations should be made upon living Indian species.   In *Dendrobium* Mr. Darwin finds that the pollinia become attached to insects visiting the flower by a viscid fluid, which is exuded by the minute rostellum immediately under the anther, when the projecting lip of the anther is pushed up by the retreat of the insect from the short nectary.

The very large Orchid Family is widely spread over the globe.   Most of the species with large showy flowers are confined to tropical countries, and grow, not upon the

ground, but upon the trunks of trees. They do not, however, attach themselves to the tree upon which they grow ;
they are not *parasites* preying upon its juices, like the
Mistletoe and Loranths, referred to at p. 223. Such plants
are distinguished as *epiphytes*. They throw out cord-like
adventitious roots freely, and the lower joints of their stems,
in many genera, become much thickened and fleshy, so as
to resemble bulbs, suggesting the name *pseudo-bulbs*, which
is specially applied to them.

Very few species of this great Family are of any economic
importance, though large numbers are prime favourites with
cultivators in Europe, from the beauty and singularity of
their flowers.

As representing a distinct Tribe of the Orchid Family,
take any species of Ladies' Slipper (*Cypripedium*), characterised by two anthers, one on each side of a shield-like
central disk, regarded as a rudimentary
anther (corresponding to the single
anther which is present in all other
Orchids). The column in the Ladies'
Slipper projects over the opening into a
large, slipper-shaped, hollow labellum.
The pollen-grains, unlike those in the
Orchids described above, are not connected together into pollinia, but they
are coated with a viscid fluid, which enables the grains to adhere when rubbed
against the stigma, which in *Cypripedium* is not viscid as
in other Orchids.

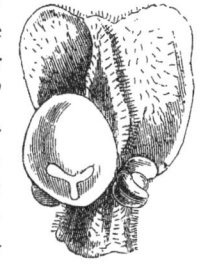

FIG. 208. Column of
Ladies' Slipper.

A small Family, the Apostasieæ, nearly related to the
Orchids, is confined to India. It serves as a link con-
necting the anomalous structure of Orchids with that of
Monocotyledons generally, differing from the former in

having two or three anthers almost quite free from the style and stigma. The flowers, moreover, are regular.

108. Natural Order, *Burmanniaceæ.*—The Burmannia Family.

Herbs with grass- or scale-like leaves. Stamens three or six. Ovary inferior.

TYPE—*Burmannia distachya.*

A small, slender, erect herb, with narrow radical and cauline ensiform leaves, terminating in a forked cyme, each branch bearing several unilateral, pretty, pale-blue regular flowers, with three-winged inferior ovaries.

Organ.	No.	Cohesion.	Adhesion.
Perianth. *leaves.*	6	Gamophyllous.	Superior.
Stamens.	3	Triandrous.	Epiphyllous.
Pistil. *carpels.*	3	Syncarpous.	Inferior.

Some of the species of this small Family are scaly, pale-coloured parasites.

109. Natural Order, *Scitamineæ.*—The Ginger and Arrowroot Family.

Herbs with irregular flowers and one free stamen (except in *Musa*). Ovary inferior.

TYPE—*Alpinia nutans* (Púnagchampa, *Beng.*).

A tall (garden) herb with lanceolate, distichous, sheathing leaves, and a terminal racemose panicle of beautiful orange and red irregular flowers.

FIG. 209. *Alpinia nutans*, much reduced.

Organ.	No.	Cohesion.	Adhesion.
Perianth. *leaves.*	6	Gamophyllous. (in two series.)	Superior.
Stamen.*	1	Monandrous.	Epiphyllous.
Pistil. *carpels.*	3	Syncarpous.	Inferior.
* Outer staminodium a petaloid "labellum."			

Sub-type—Indian Shot (*Canna indica*) differs from the Type-species principally in having three petaloid staminodia within the perianth, and the anther one-celled, the cell being upon the margin of a petaloid stamen.

Sub-type—Banana (*Musa sapientum*), with a perianth in two segments, and five fertile stamens with two-celled anthers, and a sixth stamen abortive.

The anomalous character of this curious and important Family, rich in species with beautiful flowers and affording aromatic products, is chiefly due to the petaloid development of two or more leaves of the flower, which in most other Monocotyledons are anther-bearing stamens. This makes the comprehension of the structure of these flowers difficult to beginners.

Normally, the flowers of *Scitamineæ* possess a six-leaved perianth in two series, of three each. As the ovary is always inferior, the perianth is usually regarded as gamophyllous. The six lobes, especially the three inner ones, are often unequal in form and size. Within the perianth there are six "leaves" belonging to the staminal series. Of these five are anther-bearing in Sub-type Banana, and but one in *Alpinia* and Sub-type Indian Shot. In Indian Shot the tendency to suppression of the anthers is carried to an extreme, there being but half an anther developed in the single perfect stamen of each flower. The staminal leaves which do not bear anthers are called *staminodia*. Frequently some of the staminodia are almost or altogether suppressed.

In the *Alpinia* employed as Type-species there is but one staminodium developed of the outer whorl of three staminal leaves, and it is petaloid and larger than any of the perianth-segments, forming the beautiful orange and crimson *labellum* of the flower. The remaining two staminodia of the outer whorl are to be found in the Alpinias as short teeth or lobes at each side of the labellum (see *Alpinia Galanghas*). In the beautiful garden *Hedychium* the three outer petaloid staminodia are nearly equal, so that the flower is almost regular. Of the three inner staminal leaves in the Type-species, one is developed as an anther-bearing stamen, while the other two are minute staminodia. They may be easily found on tearing the tube of the perianth open to its base, nestled around the base of the style. Sometimes they are coherent, and form a sheath around it.

Observe the entire, parallel-veined, sheathing leaves; in the Banana and Plantain of enormous size, their sheathing petioles forming a stem often several yards in height: the two-celled anther of *Alpinia* and its allies clasping the upper part of the style, the stigma projecting beyond the cells of the anther: the crest, often bifid, of the anther in *Amomum*, *Costus*, and some other genera : the pulpy aril enveloping the seeds in many genera; the embryo separated from the white, flowery, radiating albumen, by the membrane of the embryo-sac (in the genera with two-celled anthers).

The three most useful species of this tropical Family represent respectively the three Tribes indicated above. They are the Ginger and Arrowroot plants, and the Banana.

Ginger is the dried rhizome of *Zingiber officinale*. Aromatic properties more or less like those of Ginger mark the rhizomes of several species : amongst others, of the Indian *Alpinia Galanghas*, of *Costus*, and of the two species of *Curcuma*, affording Zedoary and Turmeric.

Arrowroot is the starch obtained from the tuberous rhizome of *Maranta arundinacea*. A farina of similar quality is afforded in India by the tubers of some native species of *Curcuma*.

The fruit of the Banana (*Musa*) is familiar to every resident in tropical countries.

The seeds of many Scitamineæ are aromatic and often very pungent, as Cardamoms, the product of an Indian species of *Elettaria* (*E. Cardamomum*), and grains of Paradise, afforded by a West African *Amomum.* The stalks and leaves of some Phryniums, Marantas, and allied genera, split up into narrow strips, make excellent matting, and from the leaves of one or two species of *Musa*, especially *M. textilis*, a very tenacious fibre is obtained in the Philippines, known as Manila Hemp.

A fragment of a leaf of Banana is well adapted for exhibiting spiral vessels under the microscope. There are often a number of fibres to each coil.

110. Natural Order, *Amaryllidaceæ.*—The Amaryllis Family.

Herbs, with a six-leaved petaloid perianth, six stamens, and inferior three-celled ovary.

TYPE—*Crinum asiaticum.*

A large, bulbous herb, common in gardens, with long, smooth, radical leaves, and large umbels of regular, white flowers.

Organ.	No.	Cohesion.	Adhesion.
Perianth. *leaves.*	6	Gamophyllous.	Superior.
Stamens.	6	Hexandrous.	Epiphyllous.
Pistil. *carpels.*	3	Syncarpous.	Inferior.

In this species the bulb is often prolonged above the surface of the ground so as to resemble a short trunk. In

Pancratium, and some other genera cultivated in Indian gardens, the filaments are united at the base by a membranous cup, and in the Daffodils and Narcissus of Europe a cup-like expansion called the *corona* is inserted in the mouth of the perianth-tube.

Fig. 210. *Crinum asiaticum*, much reduced.

The Family includes many very ornamental species prized in gardens. One of its most useful members is the so-called American Aloe (*Agave*), which is not an *Aloe* at all, though very similar in habit to some species of that genus. It is a Mexican plant, and its sap affords to the natives a

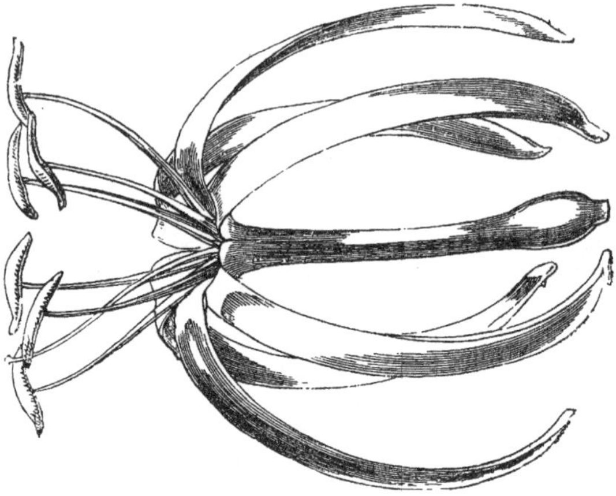

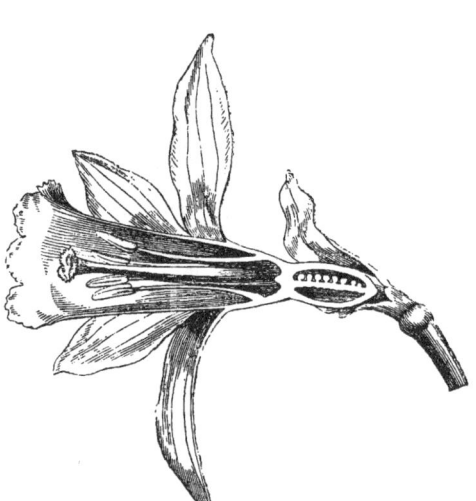

FIG. 211. Detached flower of *Crinum asiaticum.*

FIG. 212. Vertical section of flower of Daffodil Narcissus.

favourite beverage called "pulque." It is now generally introduced throughout India.

The inferior ovary is the principal mark by which this Family is distinguished from the Lilies.

111. Natural Order, *Iridaceæ.*—The Iris Family.

Herbs, with six-leaved petaloid perianth, three stamens, and three-celled inferior ovary.

TYPE—*Pardanthus chinensis.*

A common garden herb, with equitant, vertically flattened leaves, and a terminal panicle of showy, orange-coloured, regular flowers, spotted with scarlet.

Organ.	*No.*	*Cohesion.*	*Adhesion.*
Perianth. *leaves.*	6	Gamophyllous.	Superior.
Stamens.	3	Triandrous.	Epiphyllous.
Pistil. *carpels.*	3	Syncarpous.	Inferior.

A large temperate and South African Family, unimportant in India and in the tropics generally, excepting a few species planted for the sake of their showy flowers.

Our Type-species is a Chinese and Formosan plant, very common in Indian gardens.

OBSERVE the sheathing, vertically flattened leaves, arranged alternately on opposite sides of the stem, characteristic of the Family : the stigmas in Iris dilated and petaloid ; less so in *Pardanthus.*

112. Natural Order, *Hydrocharideæ.*—The Frogbit Family.

Submerged or floating plants. Flowers usually unisexual. Ovary inferior.

TYPE—*Hydrilla verticillata.*

A submerged water-plant, with small, verticillate leaves, and axillary, unisexual flowers.

Organ.	No.	Cohesion.	Adhesion.
Perianth. *leaves.*	6	Gamophyllous.	Superior.
♂ Stamens.	3	Triandrous.	Epiphyllous.
♀ Pistil. *carpels.*	3	Syncarpous.	Inferior.

OBSERVE the male flowers, which break off from the plant at the time of expansion, and float to the surface in order to fertilise the females, as in *Vallisneria,* another genus of the Family, occurring both in India and Europe. In *Vallisneria* the female flower is borne upon a long, spirally-twisted peduncle, which permits it to reach the surface while still attached.

The Type-species is one of the common water-plants used by sugar-refiners in claying sugar. The moisture which it contains slowly percolates the sugar, carrying off impurities, which are deposited in the clay. It closely resembles a species (*Elodea canadensis*) which has of late years been imported from America into England and Europe, where it has multiplied so rapidly as to obstruct navigation in still-flowing rivers and canals.

Ottelia alismoides, with radical, petiolate, ovate leaves, is common in India.

The submerged leaves of several species of this Family are well suited to show the rotation of the cell-sap in their individual cells. To observe it, place thin, longitudinal sections, or the membranous margin of a leaf, under a high magnifying power.

113. Natural Order, *Alismaceæ.*—The Water-Plantain Family.

Aquatic plants. Perianth inferior, six-leaved, three inner leaves petaloid. Pistil apocarpous.

Type—*Sagittaria cordifolia.*

Organ.	No.	Cohesion.	Adhesion.
Perianth. *leaves.*	6	Polyphyllous.	Inferior.
♂ Stamens.	6-(10)	Hex-decandrous.	Hypogynous.
♀ Pistil. *carpels.*	∞	Apocarpous.	Superior.

Fig. 213. *Sagittaria sagittæfolia,* much reduced.

A common annual weed in swamps and rice-fields, with radical, cordate leaves, and erect scapes, bearing one to three loose verticils of small, white, polygamous flowers.

The lower flowers of the inflorescence are usually female. In the allied *S. sagittæfolia,* with arrow-head leaves, the male flowers have numerous stamens.

FIG. 214. Longitudinal section of
achene of *Alisma.*

FIG. 215. Embryo of same removed
from the seed.

The Water Plantains represent the Ranunculus Family amongst Monocotyledons, having the pistil nearly or wholly apocarpous, and hypogynous stamens. They differ widely in their embryo, as well as in other points of structure.

114. Natural Order, *Naiadaceæ.*—The Pondweed Family.

Floating or submerged plants. Perianth o, or four-leaved. Pistil apocarpous.

TYPE—Common Pondweed (*Potamogeton natans*).

Organ.	No.	Cohesion.	Adhesion.
Perianth. *leaves.*	4	Polyphyllous.	Inferior.
Stamens.	4	Tetrandrous.	Hypogynous.
Pistil. *carpels.*	4	Apocarpous.	Superior.

An aquatic herb, with floating, opaque, oblong or elliptical, stipulate leaves, and pedunculate spikes of minute, greenish, hermaphrodite flowers.

OBSERVE the seeds, like those of most Monocotyledons of aquatic habit, destitute of albumen.

GLUMIFERÆ.

115. Natural Order, *Cyperaceæ.*—The Sedge Family.

Grass-like herbs. Sheaths of leaves not split. Scale next the flower with a *median* nerve.

TYPE—*Cyperus Iria* (or any other species of *Cyperus*).

A grass-like plant, one to two feet high, with an angular, solid stem, closed leaf-sheaths, and irregularly umbellate, distichous spikelets of hermaphrodite flowers singly borne in the axils of imbricating glumes.

Organ.	*No.*	*Cohesion.*	*Adhesion.*
Perianth.	o	...	...
Stamens.	3	Triandrous.	Hypogynous
Pistil. carpels.	3	Syncarpous.	Superior.

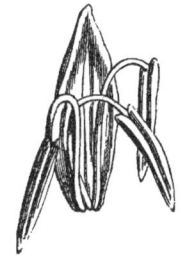

FIG. 216. Flower of Cyperus with subtending glume ; enlarged.

FIG. 217. Male flower of Carex.

A large Family, represented in every quarter of the globe by numérous species, generally abounding in wet places. The sedges resemble grasses in habit, but may be distinguished by their usually solid and angular stem, the closed (not split) sheaths of their leaves, and the flowers each borne in the axil of a single bract (*glume*), without the additional pale of grasses.

FIG. 218. Female flower of Carex.   FIG. 219. Vertical section of fruit of same.

In the largest genus of the Family *Carex*, and in *Scleria*, of both of which genera there are numerous Indian species, the flowers are unisexual. In the latter genus the nut is often white, shining, and very hard and bony.

In the Type-species the flowers are arranged in distichous spikelets, but in most of the Indian genera they are in the axils of glumes which are regularly imbricated all round the spikelets.

In several genera, as *Scirpus*, *Fuirena*, and *Rhynchospora*, a perianth is represented by from three to six hypogynous bristles or scales.

The rhizomes or small tubers of a few species of the Family are used in native medicine, and the stems and leaves of others (as *Cyperus distans*) are employed to make coarse matting and cordage. The famous *Papyrus*, used as paper by the ancient Egyptians, was prepared from the pith-like tissue of the tall stems of a plant nearly allied to the Type-species. It is now nearly or quite extinct on the banks of the river Nile.

Some species, with far-spreading rhizomes, serve a useful purpose in binding together the shifting sands of the coast and river-banks.

116. Natural Order, *Gramineæ.*—The Grass Family.

Herbs (except Bamboos). Sheaths of leaves usually split in front. Scale (*pale*) next the flower without a *median* nerve.

TYPE—Wheat (*Triticum vulgare*).

A generally cultivated, erect annual, with an unbranched, jointed, hollow, leafy stem (*culm*), the leaf-sheaths split in front, and distichous spikelets of flowers.

FIG. 220. Dissected spikelet of Wheat (*Triticum vulgare*).

Organ.	No.	Cohesion.	Adhesion.
Lodicules.	2	(Diphyllous.)	Hypogynous.
Stamens.	3	Triandrous.	Hypogynous.
Pistil. carpels.	2	Syncarpous.	Superior.

The structure of the flower of Wheat has been already described (p. 55).

The following list shows in what particulars several of the more frequent and more important genera of Indian grasses deviate from Wheat in the form of their inflorescence and structure of their flowers.

  * *If empty glumes or imperfect flowers be present in the spikelets, they are inserted below the single hermaphrodite flower (Tribe, PANICACEÆ).*

Oryza (RICE).—Panicle. Spikelets one-flowered, pedicellate. Outer glumes minute. Stamens six.

Zea (MAIZE, or INDIAN CORN).—Monœcious. The male flowers in terminal, panicled racemes; spikelets pedicellate. The female flowers sessile, in lateral, erect spikes.

Coix (JOB'S TEARS).—Monœcious. The male flowers loosely spicate; the female enclosed in a small involucre, which becomes at length of stony hardness.

Alopecurus.—Spicate. Spikelets nearly or quite sessile, one-flowered. No pale or lodicules.

Paspalum.—Several digitate spikes. Spikelets sessile, unilateral, with one perfect flower.

Panicum.—Variously panicled: in section *Digitaria*, of several digitate branches; in *Setaria*, cylindrical and spicate. Each spikelet contains one perfect flower, and there are three empty glumes below it.

Penicillaria.—Cylindrical, spicate panicle. Spikelets with one perfect flower, without lodicules.

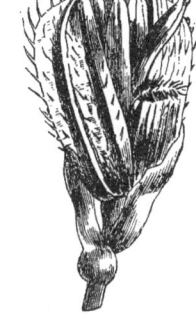

Fɪɢ. 221. Rice (*Oryza sativa*), much reduced.

Fɪɢ. 222. One-flowered spikelet of Rice.

Saccharum (Sᴜɢᴀʀ-ᴄᴀɴᴇ).—Panicle. Spikelets in pairs, one sessile, one pedicellate, with one perfect flower, and three empty glumes below.

Andropogon (Kᴜs-Kᴜs) and *Sorghum.*— Paniculate. Spikelets in pairs, one sessile, the other pedicellate. Sessile spikelet with one perfect flower, and three empty glumes. The pale is sometimes wanting.

 ** *If empty glumes or imperfect flowers be present in the spikelets, they are inserted above the hermaphrodite flower or flowers* (Tribe, Pᴏᴀᴄᴇᴁ).

Sporobolus.—Long, slender panicle. Spikelets pedicellate, minute, one-flowered.

z

Arundo (Donax).—Panicle.   Spikelets pedicelled, two-
to four-flowered.

Cynodon (Dactylon).—Digitate spikes.   Spikelets sessile,
one-flowered.

Eleusine (Coracana and *indica).*—Spikes digitate.   Spikelets sessile, unilateral, two- to six-flowered.

Fig. 223. Spikelet, dissected, of Vernal Grass (*Anthoxanthum*).   The lowest pair of
scales, right and left, are the outer glumes ; then come two awned empty glumes ;
then, to the right, the flowering glume, and to the left the small pale ; lastly the
two stamens and the pistil.

Hordeum (BARLEY). — Distichous, spicate.   Spikelets
sessile, ternate, each one-flowered.   The lateral spikelets
imperfect (in two-rowed Barley) or perfect (in six-rowed
Barley).

Bambusa (BAMBOO). — Arborescent.   Spikelets sessile,
clustered, or verticillate, with several perfect flowers.
Stamens six.   Lodicules three.

OBSERVE the stem, called a *culm,* usually hollow (*fistular*),

with a few joints below, and sheathing leaves : the sheath
of the leaf split down in front : the scale, called a *ligule*,
at the base of the blade of the leaf, where it is given off
from the sheath ; it is usually membranous, short or long,
blunt or acute : the fruit commonly regarded as a seed, and
technically distinguished as a *Caryopsis*. It consists of a
thin pericarp closely adhering to the solitary seed. It often

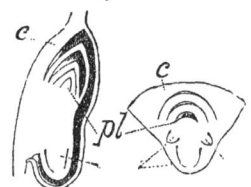

FIG. 224. Longitudinal section through
a grain of Wheat, showing the
oblique embryo at the base of copious
albumen.

FIG. 225. Longitudinal and transverse
sections of the embryo of Wheat :
c cotyledon ; *pl* plumule ; *r* radicle.

happens that the pale becomes adherent to the caryopsis
after flowering, and sometimes, also, the flowering glume.
When this is the case in corn-producing species, it is
removed by grinding, as in the case of Barley and Oats.
In Wheat and Indian Corn the caryopsis is free, that is,
it is not adherent to the pale.

The Grass Family includes probably from 3,000 to 4,000
species. Amongst these are several affording large farinaceous seeds, distinguished as Cereals or Corn-grasses,
which are of primary importance to the human race, and
have been cultivated from the remotest antiquity.

Indian genera, including food-producing species, are
printed in small capitals in the above list. To these may
be added the Oat (*Avena*) and Rye (*Secale*) of cool
countries.

Rice is stated to furnish a larger proportion of food than
any other single species.

Maize, or Indian Corn, the largest Cereal, is a plant of American origin, now extensively cultivated in hot countries. As noted above, the flowers are monœcious; an exceptional character in the Grass Family.

As is usual in plants which have been under cultivation for a very long period, most of the cereal grasses occur under numerous slightly different forms or varieties, which have probably originated by cultivation under various conditions of soil and climate, and the selection of sorts, best suited for particular purposes, by man.

Apart from their importance as Cereals, Grasses are invaluable on other grounds. Some, as the Sugar-cane (*Saccharum officinarum*), abound in a saccharine juice. From this species the bulk of the sugar of commerce is prepared. Others, as some species of *Andropogon* (Lemongrass, Kus-kus), furnish fragrant essential oils. The culms of grasses (straw) are largely used in matting, plaited work, and thatching. Many are fodder-grasses. *Cynodon Dactylon* alone is said to furnish three-fourths of the food of horses and cattle in India.

The applications of the rapidly-growing woody stems and leaves of the Bamboo by the natives of India and Eastern Asia are innumerable. A small room in the Kew Museum is occupied by products and manufactures of the Bamboo, including umbrella, chair, walking-stick, tiger-trap, bow and arrow, window-blinds, bowls, combs, musical instruments, cloth, paper, &c. &c.

CHAPTER IV.

Thus far I have avoided reference to those plants which are commonly regarded as Flowerless, and which have long been classed together under the general term of *Cryptogams*, from the apparent absence of organs corresponding to the stamens and pistil of the plants which have hitherto occupied our attention.

We have passed these plants by because, from the considerable difference which obtains between their structure (both of the Reproductive and of the Nutritive organs) and that of Flowering Plants, they cannot be conveniently studied together. Any study, however, of the Vegetable Kingdom from which they are wholly excluded must be exceedingly incomplete; and now that facility in observing has been acquired, attention may be directed to these so-called lower plants, with a fair chance of comprehending the relation in which they stand to the Flowering Plants already familiar to us, and of mastering a few of the principal features of their leading Families.

The more logical course might seem to be to study first these simple forms, and progress from them to the more complicated, to which latter we have hitherto confined our attention; but from the excessive minuteness of their essential organs, and our imperfect acquaintance with many details of their structure and function, it is practically the best plan to leave them to the last, in a course of Elementary Botany like the present. Space compels me to be brief in describing Cryptogams; and those who desire to extend their acquaintance with them I must refer to special works which treat of them in detail. A few of these are noticed in the list of works on Indian Botany given in the Appendix.

All the plants which I have described in foregoing chapters produce seeds containing an embryo, provided (with unimportant exceptions) with one or more rudimentary leaves, which we have termed Cotyledons. The plants which we have now to consider do not produce a seed containing an embryo, but are multiplied by minute reproductive bodies, called *spores*.

The spores consist usually of a simple cell. As they are destitute of an embryo, there cannot, of course, be any distinction of radicle, plumule, and cotyledons; in consequence of the absence of the latter, Flowerless Plants are termed *Acotyledons*.

Although we speak of these plants as *Flowerless*, it must be borne in mind that they do possess organs analogous to those which are essential to the flower, but they are so disguised, and often so simple, that they have been in many cases but recently recognised as such.

The more important Families of Acotyledons (Cryptogams or Flowerless Plants) are—

Possessing distinct stem and leaves .
{ Ferns (*Filices*).
Club-mosses (*Lycopodiaceæ*).
Horsetails (*Equisetaceæ*).
Mosses (*Musci*). }

No distinction between stem and leaf
{ Mushrooms and Moulds (*Fungi*).
Lichens (*Lichenes*).
Sea-weeds (*Algæ*). }

1. Natural Order, *Filices.*—The Fern Family.

Leaves (*fronds*) curled upon themselves like a crosier (*circinate*) before expansion. Fructification upon the under surface of the frond, consisting of minute usually densely-clustered capsules (sporanges) of one kind, containing microscopic doubly-coated cells (spores) destitute of an embryo, but capable of developing a small green leafy expansion (prothallus) bearing the essential Organs of Reproduction.

TYPE.—*Aspidium (Nephrodium) molle.*

FIG. 226. *Aspidium (Nephrodium) molle*, much reduced.

A perennial herb with a short ascending rhizome, bearing a terminal tuft of large, annual, broadly lanceolate, thinly pilose, bipinnate fronds, two to four feet in length.

Upon the under surface of the fertile fronds the fructification is arranged in small globose clusters, which are
brownish when ripe.   These clusters are called *sori*

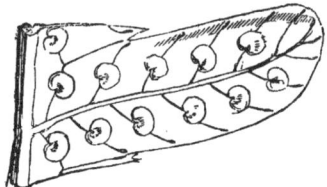

FIG. 227. Pinnule of *Aspidium molle*, with a double row of reniform sori.

(each cluster a *sorus*), and there are usually two rows
of the sori upon, at least, the lower lobes (*pinnules*) of
each of the *pinnæ* of the frond. Occasionally only the
lower sori of each pinnule are developed, and then they
occur in a single row on each side of the midrib of the
pinnæ.   The sori are protected at first by a peltate membrane (*indusium*), which at length withers up, exposing the
minute-stalked *sporanges* of which each sorus is composed.

The sporanges require examination with a magnifying-
glass.   They will be found to be capsules opening transversely, with a vertical elastic ring up the side and over
the top, which serves as a hinge.   They each contain an
indefinite number of spores, which are liberated on dehiscence of the sporange.

Many Indian Ferns depart considerably from our Type-
species in the form of the frond, the form and arrangement
of the sori, the absence of indusium, and in the structure
and mode of dehiscence of the sporanges.   The difference
presented in these respects by some of the commoner and
more remarkable Indian genera are noted in the following
list.

* *Sporanges distinct (not cohering), and provided with*
    a more or less distinct ring (annulus).

(*a.*) *Annulus vertical.*

Polypodium.—Sori on the under surface of the frond,
nearly or quite round. Indusium o.

Niphobolus (adnascens).—Sterile and fertile fronds distinct.
The upper part of the linear fertile frond crowded with
pedicellate sporanges between the midrib and margins,
mixed with stellate, peltate scales.

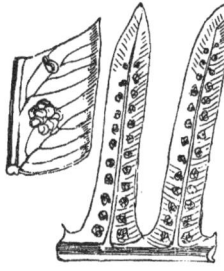 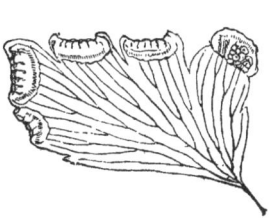

FIG. 228. Pinnules and sori of *Polypodium.*   FIG. 229. Pinnule of *Adiantum.*

Adiantum (Capillus-Veneris, or *caudatum).*—Sori on the
margins of the pinnules, covered by the indusium, which is
attached by its outer edge, opening (free) within.

Pteris.—Sori in a continuous line upon the margin of the
pinnules. Indusium continuous, attached, as in *Adiantum,*
by its outer edge.

Blechnum (orientale).—Sori in continuous lines on each
side of, and parallel with, the midrib of the frond or pinnule.
Indusium opening along its inner edge.

Asplenium.—Sori scattered on the under surface ; not
marginal, and generally oblique to the midrib of the frond
or pinnule. Indusium membranous.

Aspidium (*molle*).—Sori nearly or quite globose on the under surface of the frond. Indusium peltate (in § *Aspidium* proper), or reniform and attached by a point at the side (in § *Nephrodium*, to which belongs our Type-species).

FIG. 230. Pinnule of *Nephrodium.*

Lindsæa (*ensifolia*).—Sori continuous along the margin of the segments, with a narrow continuous indusium opening on the outer edge.

Davallia (*polypodioides*).—Sori terminating veins, nearly or quite on the margin of the pinnules. Indusium cup-shaped, adhering to the pinnule and opening towards the margin.

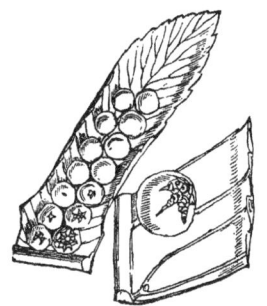

FIG. 231. Pinnule and sorus of *Cyathea.*

Cyathea (*spinulosa*). — Arborescent; sporanges forming globose sori in the axils of forking veins, contained in a cup-shaped indusium, which opens at the top by a few teeth.

Alsophila (glabra). — Arborescent, sometimes fifty feet high. Sori globose, without an indusium.

(*b.*) *Annulus oblique, transverse, incomplete or cap-like.*

Ceratopteris.—An aquatic Fern, with distinct sterile and fertile fronds. Sori continuous along the veins of the narrow lobes of the fertile fronds. Indusium formed by the revolute margins of the lobes. Annulus of the sporanges incomplete. Spores marked with three separate series of concentric rings.

Hymenophyllum.—Sori at the end of veins, terminating lobes of the frond. Sporanges with a horizontal annulus, sessile upon a slender column within a two-valved indusium. The fronds of this genus and the following are half-pellucid. They are often very small and the rhizome is densely matted.

 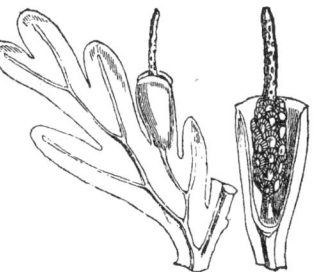

Fig. 232. Pinnule and sorus of   Fig. 233. Pinnule and sorus of
        Hymenophyllum.   *Trichomanes.*

Trichomanes.—Sori and sporanges as in *Hymenophyllum.* The indusium tubular, and the hair-like column projecting beyond it, the sporanges clustered near its base.

Gleichenia (dichotoma). — Sori scattered, of few sessile sporanges. Sporanges opening vertically with a transverse ring. Indusium o.

Lygodium.—Slender, climbing Ferns with pinnate fronds. Sporanges on the under side of marginal or terminal lobes of the pinnules, singly attached by the middle, each under an imbricating bract-like indusium. Sporanges with longitudinal striæ.

Osmunda (regalis). — Sporanges densely covering the upper segments or pinnules of the fertile fronds, so as to recall a panicled inflorescence. Annulus o.

        ** *Sporanges cohering. Annulus* o.

Kaulfussia.—Sporanges radiating, forming round, concave sori, opening by slits at the top.

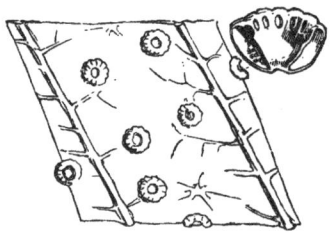

FIG. 234. Portion of frond and sorus of *Kaulfussia.*

      *** *Fronds of two distinct parts—sterile or leafy, and fertile or sporange-bearing.*

Ophioglossum.—Sporanges on an undivided spike.
Botrychium.—Sporanges on a divided or pinnatifid spike.

The development of young Ferns from their spores may be watched by growing the spores upon damp soil covered by a bell-glass. From the germinating spore arises a small, green, leafy expansion, termed the *prothallus,* which gives off from its under surface numerous delicate root-fibres. Scattered amongst these fibres, especially on the thicker part of the prothallus, are several microscopic cellular

bodies of two distinct kinds. One kind, the more numerous, called *antheridia*, contain a number of extremely
small vesicles, each of which liberates a spirally-twisted
filament, called an *antherozoid*, which performs the function
of a pollen-grain. The other kind, called *archegonia*, contains an embryonal cell which, fertilised by the antherozoids
set free by the antheridia, gives origin to a bud which
gradually developes into an independent Fern-plant.

The antheridia and archegonia require a high magnifying
power for their examination. Their true nature has been
understood only within the last twenty years.

The stem of Ferns differs from that of both Dicotyledons
and Monocotyledons in its growing solely by additions to
the summit, below which all the tissues are completed with
their first formation, and in the arrangement of the fibro-
vascular bundles, which form an interrupted circle around
a cellular axis which usually decays away, so that old stems
become hollow. From the mode of growth of Fern-stems,
by additions to the extremity, they have been termed
Acrogens (point-growers). Some of the Himalayan species
of *Alsophila* and *Cyathea* (Tree Ferns) form tall woody
stems, well adapted to illustrate this structure.

2. Natural Order, *Lycopodiaceæ.*—The Club-moss Family.

Low, trailing, or tufted, usually slender, wiry herbs, with
small, two- (four-) rowed or scattered imbricating leaves.
Fructification consisting of sporanges in the axils of the
stem-leaves or collected in terminal bracteate spikes, containing spores of one of two kinds, either minute and
indefinite, called *microspores*, developing antheridia, or larger
and definite, called *macrospores*, developing a prothallus
bearing archegonia.

Lycopodiaceæ generally affect a humid climate, and in

India they are chiefly hill-plants.   One of the commoner
Indian species is the Stag's-horn Club-moss (*Lycopodium
clavatum*) of British moorlands.   It has a very wide geographical distribution, occurring in the Southern hemisphere,
as well as through Northern Asia and Europe and in North
America.   The Indian form, represented in the woodcut,
differs in trivial characters from the European type, but it
may be regarded as specifically identical.

FIG. 235. *Lycopodium clavatum,* Indian form, reduced.

In SUB-TYPE I (*Lycopodium*) the leaves are usually subulate
and imbricated around the stem.   Sporanges containing

definite spores (macrospores) have not been observed in the genus.

In Sub-type 2 (*Selaginella*) the leaves are of two kinds, the larger usually obliquely oblong or ovate, distichously

Fig. 236. *Selaginella caulescens*, reduced.

arranged in the plane of ramification, the smaller stipule-like, appressed, and intermediate. The sporanges are of two kinds, containing either macrospores or microspores.

In Sub-type 3 (*Psilotum*) the leaves are very minute, distant, and scale-like, and the sporanges three-celled.

The genus *Isoetes*, represented in India by two aquatic species, very nearly related to a British one, differs remarkably in habit from Club-mosses, but agrees in having two kinds of sporanges as in *Selaginella*. The species are either aquatic or terrestrial, with a short unbranched stock, and tufted linear or subulate sheathing leaves, in the bases of which the sporanges are embedded. The mode of reproduction in *Lycopodium* is not yet cleared up, but in *Selaginella*, in which the lower sporanges contain macrospores and the upper microspores, a narrow scarcely protruding prothallus is developed upon the former, bearing archegonia upon its surface, which are fertilised by antherozoids set free by the microspores.

3. Natural Order, *Equisetaceæ.*—The Horsetail Family.

Herbs with hollow jointed stems, with or without slender whorled jointed branches. Fructification, a terminal spike, consisting of numerous closely-packed peltate scales bearing

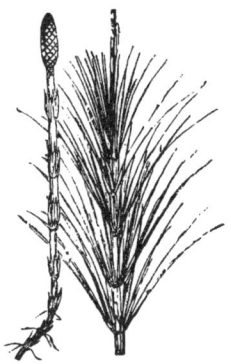

Fig. 237. Unbranched fertile and branched barren fronds of Horsetail.

sporanges of one kind around their margins, parallel with their short stalks. Outer coat of the spores splitting into elastic attached hygroscopic filaments (elaters). Developing a prothallus as in Ferns.

The mode of reproduction of *Equisetaceæ* is similar to that of Ferns.

In *Equisetum diffusum*, the commonest species of the Indian mountains, very nearly related to a common European species (*E. palustre*), the fertile or sporange-bearing fronds are either simple or branched.

In *E. hyemale*, an unbranched species of Northern Europe, the epidermis contains so much silica that bunches of the stems are sold for polishing metal.

4. Natural Order, *Musci.*—The Moss Family.

Minute herbs with filiform or slender wiry stems, and alternate usually spirally-arranged and imbricating leaves; destitute of vascular tissue. Fructification consisting of a

FIG. 238. Hair-Moss (*Polytrichum*). *a* seta bearing a sporange ; *b* sporange covered by its calyptra ; *c* head of antheridia, forming the male inflorescence.

stalked sporange, usually with a central axis containing microscopic double-coated spores of one kind, capable of developing a thread-like branching filament, upon which

leafy shoots give origin to new Moss-plants, which bear
the reproductive organs when fully developed.

While in Ferns the spores give origin on their germination to a minute temporary organ, upon which antheridia
and archegonia are developed, in Mosses the complete
vegetative system—that is, roots, branches, and leaves—
is developed from the spores without the intervention of
sexual organs. Upon the leafy branches antheridia and
archegonia, analogous, though different in structure, to
those of Ferns, are developed. From the archegonia,
fertilised by the spiral antherozoids liberated by the antheridia, arise the sporanges, usually borne up by a slender
peduncle, the *seta,* and capped by the upper portion of
the archegonium, which is torn away by the rising sporange,
for which it forms the calyptra. The mouth of the
sporange is closed until mature by a lid (operculum), which
separates when ripe, exposing a row of minute teeth around
the margin of the sporange, in many Mosses. These teeth
form the peristome.

In Mosses, vessels are wholly absent from both stem
and leaves; hence they—together with the plants grouped
under the three following Families, all of which are destitute
of vascular tissue—are termed Cellular Plants. Nearly all
the plants which we have hitherto noticed, whether of
Flowerless or Flowering Families, contain vessels, and are
consequently termed Vascular Plants.

The species of Musci are very numerous, especially in
cool and cold climates. In India they are almost confined
to the Himalaya, the mountains of Eastern Bengal, the
Peninsula, Ceylon, and the Malayan chain.

5. Natural Order, *Fungi.*—The Mushroom and Mould Family.

TYPE—Any species of Agaric (*Agaricus*).

With a vegetative system growing under the surface of soil containing decaying organic matter, and consisting of a flocculent network of delicate cellular threads, forming what is called the *mycelium.* The fructification is borne above the surface, in the form of an umbrella-like disk called the *pileus*, upon a stout stem. The margin of the pileus is at first united by a membrane to the stalk, from which it breaks away, leaving a ring-like scar. Upon the under-side

FIG. 239. Mushroom (*Agaricus*).

of the pileus numerous vertical plates radiate from the top of the stem to the margin of the pileus. If a very thin, transverse section of one of these plates be cut with a sharp knife, and examined under a powerful microscope, the surface will be found to be studded with large cells, each of which bears four very minute stalked spores upon its apex.

A A 2

Other Fungi depart very widely from this Type, but nearly all agree in the absence of green colouring-matter and of starch in their cells, and in their dependence upon decaying animal or vegetable matter for support. They are mostly short-lived, and often deliquesce when mature, though some, as the Touchwoods (*Polyporus*), are hard and woody.

In many Fungi there is no distinction of stem and pileus; and the spore-bearing cells clothe excavations in the cellular substance of the Fungus (as in Puff-balls, *Lycoperdon*), or the spores may be formed in the interior of certain cells called *asci*, two, four, or more together, as in the subterranean esculent European fungus called Truffle (*Tuber*).

Some botanists divide the Fungi into two Tribes : (1) with the spores borne upon the exterior of cells called *basidia*, and (2) with the spores developed in the interior of cells called *asci;* those of the former Tribe being termed *Sporiferous*, those of the latter *Sporidiferous*. The variety in arrangement of the reproductive system of the Fungi is extreme, and there are very many species which are as yet very imperfectly understood.

Though a few of the Fungi are esculent, many are dangerous, and some poisonous. None should be eaten unless perfectly sound, and species with a disagreeable odour should be avoided. Many Fungi are very injurious, destroying large quantities of agricultural produce, timber, and miscellaneous substances, when circumstances favour their development. The Wheat Mildew, Smut, and Bunt of Cereals, Ergot, Hop-blight, the Moulds, and Dry-rot, are all Fungi. The Vine and Potato diseases are also due to the ravages of minute species, which multiply with great rapidity. As their spores are excessively minute, they cannot be excluded by any mechanical contrivance.

6. Natural Order, *Lichenes.*—The Lichen Family.

Lichens occur either as crust-like or leafy expansions, or in little branching shrubby tufts, usually coloured grey, yellow, or greenish-yellow. They spread everywhere in cool climates—over stones, brick-walls, the bark of trees, and even upon the most exposed rocks of alpine and arctic countries, forming the very outposts of vegetation, and growing at the expense, almost solely, of the atmosphere and the moisture which it bears to them. In the tropics the relative proportion of lichens growing upon the leaves of trees (*epiphyllous* lichens) is large, and gives a special character to tropical Lichenology. Unlike Fungi, the Lichens are long-lived, and intermittent in their growth, being at a standstill, and often crumbling away, when the weather is dry. They differ, also, from Fungi in containing a green-coloured layer under the epidermis, consisting of cells called *gonidia*, which may be regarded as answering to the buds of higher plants, since, when set free, they develope new lichens, and thus multiply the plant. The true reproductive organs are contained in special receptacles, either exposed upon the upper surface of the lichen or buried in its tissue, the spores being contained in narrow cells similar, in some species, to the asci of Sporidiferous Fungi.

Several species, as *Lecanora* and *Roccella*, afford a valuable purple and mauve dye; and a few are edible, as the so-called Iceland Moss (*Cetraria islandica*). The Reindeer Moss (*Cladonia rangiferina*) is a lichen extremely abundant in polar regions, serving as food to the reindeer. One or two species of *Parmelia* growing upon rocks in Southern India are used in medicine.

7. Natural Order, *Algæ.*—The Sea-Weed Family.

This Family includes an enormous number of species, nearly all adapted to grow under water, though by no means all marine, as many are wholly confined to fresh water. They vary to an extraordinary extent in size, form, and mode of reproduction. Some are microscopic and individually invisible to the naked eye; whilst others, especially some marine species, attain a large size—a few, indeed, measuring some hundreds of feet in length. The simplest forms consist of single microscopic cells: hence they are called Unicellular Algæ. These multiply by division, and also by a kind of sexual reproduction, analogous to that of the higher plants, in which the contents of two distinct individual cells become commingled, and the resulting mass finally resolves itself into a number of new individual cells or plants.

Forms of a higher grade of structure are represented by the fine hair-like filaments which we find floating in rivulets and tanks, rooted at one extremity to stems or to larger water-plants. Many of these *filamentous* species (*Confervoideæ*) multiply themselves by the contents of the cells which form their filaments being resolved into innumerable minute moving bodies, called *zoospores*, which break out of the cells and rush about in the water until they finally settle down and grow.

The higher species, such as the Olive-coloured Sea-Weeds (*Fucus*), which clothe the rocks between tide-marks upon the shores of northern countries, possess a complicated reproductive system of spores and antheridia, contained in *conceptacles* embedded in the thickened extremities of the divided fronds. Their mode of reproduction, adapted to the medium in which the species grow, agrees in essentials with that which is characteristic of Ferns and Mosses; with this difference, however, that the spores themselves are directly fertilised and rendered capable of independent growth by the contact of the minute spermatozoids contained in the antheridia.

The marine species vary in colour, some being usually olive, others red or green. The colour is employed as an aid in their classification.

Many species are used for food. Some of the larger marine species used to be burnt for the sake of their alkaline ash (kelp) and for the iodine which they contain.

CHAPTER V.

SPECIMENS which are to be dried, so that they may be kept in a HERBARIUM and referred to or examined at a future time, ought not to be gathered at random, but should be selected as average representatives of their species, unless they be designed to show some departure from the typical form. If herbaceous plants, they ought, if possible, to be taken up, when in flower, by the root, and the root should be pressed, if not too large, along with the rest. If the radical leaves be withered at the time of flowering, another specimen should be gathered at an earlier season to show them, as the radical leaves are often very different in form from those of the stem. Besides expanded flowers, the bud and ripe fruit should be shown; and if these cannot be had upon a single specimen, other examples should be collected, to show the plant in its different states. A strong knife or small trowel will be found useful to dig up the specimens.

The specimens should not be allowed to wither before reaching home. They may either be carried in a tin box, or loosely spread between sheets of paper in a portfolio. Fig. 240 shows a collecting portfolio, which may be made of two pieces of pasteboard sixteen inches long by ten

inches wide, fastened, as shown in the cut, by tape or
straps. A few sheets of absorbing-paper must be kept
in the portfolio.

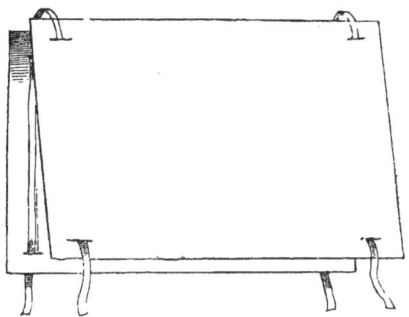

FIG. 240.   Collecting Portfolio.

In laying out specimens for the press use plenty of paper,
so that their moisture may be quickly absorbed, and the
danger of mould avoided. The specimens should be laid
between the sheets of drying paper in as natural a position
as may be, taking care not to crumple the leaves or flowers.
If the specimens be too long for the paper, they may be
carefully folded or cut in two. Delicate flowers should be
carefully folded in paper when gathered, and kept flat.
Do not arrange every specimen just in the middle of the
paper, but dispose them in such a way, that were a pile of
them in their papers raised two feet high they would not
topple over : this will equalize the pressure. Several dry
sheets ought to be laid between each layer of fresh specimens, the quantity of paper depending upon the thickness
and succulence of the plants to be pressed. In the case of
thin-leaved and delicate plants, it is not a bad plan to treat
the sheet of paper upon which the specimen is laid as part
of the specimen, removing it, every time the papers are

changed, with the specimen undisturbed upon it, to the dry
sheets.   Pasteboards, or, better still, "ventilators" (made
the size of the paper, of narrow strips of deal at short
distances apart, nailed together in two layers at right-angles
to each other, as shown in the cut, Fig. 241), may be introduced at intervals between the layers of paper until the pile
be ready for the press, which may consist simply of two
stout boards, made so that they cannot bend or warp.
Between these boards the paper and specimens must be
placed, and a weight of stones or metal, not less than
50 lbs. or 60 lbs., laid upon the top.

The papers should be changed, several times, once a day,
and then at longer intervals, until the specimens are quite
dry, when they should be removed from the press.   If fresh

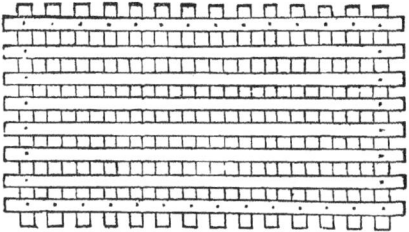

FIG. 241. Ventilator.

specimens be placed in the press, while others are in process of drying, they must be carefully separated by pasteboard, or by a thick layer of paper.   The length of time
which specimens ought to remain in the press varies with
their nature, whether dry or succulent, and with the kind
and quantity of paper used.   Common stout brown paper
answers very well.   It may be cut to any size, but, generally,
it should not be less than sixteen or eighteen inches long
by ten inches wide.   Practice will soon suggest many little

useful expedients in drying plants which it is needless should be detailed here.

The dried specimens should always be accurately labelled with the locality, name of finder, name of the plant, and any other details which may be thought desirable. They may either be kept loose in sheets of paper, or (and necessarily, if intended for use in a school, or for frequent consultation) mounted upon sheets of stout cartridge-paper of a larger size than foolscap, say about $16\frac{1}{2}$ ins. by 10 ins. A ream consists of 960 half-sheets, sufficient for as many specimens.

The specimens should be fastened to the Herbarium paper with hot glue, about the consistency of cream, the glue being laid on the specimens with a hair pencil. The newly-mounted sheets should be placed between waste paper or newspapers, and pressed overnight, before they are finally retouched and placed in the Herbarium. Straps of gummed thin paper may be fastened over the thicker parts of the specimens, to prevent them breaking loose from the paper when accidentally bent.

The mounted specimens belonging to the same genus, or a part of them if the genus be a large one, may be placed in a folded sheet of a stronger and coarser paper than that upon which the specimens are glued : upon this cover, at the bottom, the name of the genus and of its Natural Order may be marked. The Genera should be arranged in their Natural Orders, the Natural Orders in their respective Divisions and Classes, and the whole placed in a suitable cabinet, which, however, need not be procured just at first.

Whatever the form of the cabinet in which the Herbarium is kept, it should be securely closed, so as to exclude dust, and camphor should be placed upon the shelves, unless the specimens are well washed over with a preservative solution

before being laid in.*   The accompanying cut shows an
excellent form of cabinet, made of deal, similar to those in
use at the Herbarium of the Royal Gardens, Kew.

* The preservative solution may consist of corrosive sublimate dissolved in spirits of wine, in the proportion of two drachms to the pint.
It is very poisonous, and should be kept labelled, and used with care.

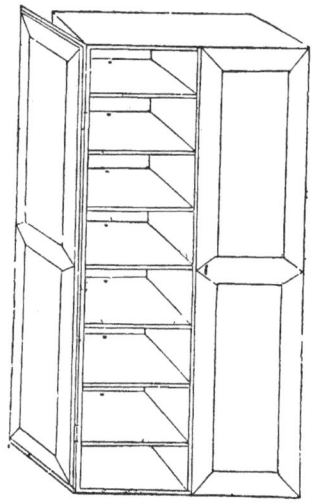

Fig. 242.   Herbarium Cabinet.

APPENDIX.

I.

HOW TO DESCRIBE PLANTS.

WHEN the student has acquired facility in filling up schedules from plants belonging to all the principal divisions of Phanerogamia, it is desirable that he should proceed to describe specimens more at length, as shown in the following examples. As the principal use of the schedules is to direct the attention to certain important points of structure, care must be taken never to omit reference to these important points in describing plants in this way. If, however, as is best, the description be headed with the CLASS and *Division* to which the plant belongs, it is not necessary (excepting in Examination exercises) to detail all the characters which are implied by referring it to such Class and Division. The organs must be described *seriatim* in the order of their development.

ORANGE : *Citrus Aurantium.*

CLASS, Dicotyledons.   *Division,* Thalamifloræ.

A wholly glabrous evergreen shrub or small tree, with shining simple (unifoliolate) leaves, fragrant white axillary flowers, and globose baccate fruits.

EXTREMITIES smooth, green, glabrous, the internodes obtusely angular above ; occasionally armed with short straight acute axillary spines.

LEAVES alternate, petiolate, unifoliolate, elliptical elliptic-oval or broadly lanceolate, acute obtuse or narrowly retuse, often broadly acuminate, obsoletely crenate-serrulate, rather coriaceous, translucently glandular-dotted, lamina articulated at the base to the petiole, which is often more or less distinctly winged.

FLOWERS in shortly pedunculate or sub-sessile few-flowered fascicles from the axils of the upper leaves, regular, hermaphrodite, white and fragrant.

CALYX inferior, cup-shaped, persistent, quinque-dentate, teeth deltoid.

COROLLA hypogynous, polypetalous, much exceeding the calyx; petals normally five (varying to eight), linear or oblong, fleshy, spreading or recurved, imbricate in æstivation.

STAMENS hypogynous, indefinite, polyadelphous; filaments compressed, variously coherent; anthers oblong, two-celled, dehiscing longitudinally.

PISTIL syncarpous; ovary superior globose, seated upon a fleshy annular or hemispherical disk, many-celled; style erect, terete, deciduous, stigma capitate lobulate; ovules indefinite, biseriate.

FRUIT globose, many-celled: pericarp fleshy, replete with minute immersed receptacles of aromatic essential oil, glabrous, rugulose; septa membranous; cells usually few-seeded (or seeds abortive), filled with a succulent cellular tissue developed from the inner wall of the pericarp.

SEEDS exalbuminous; testa coriaceous; embryo with fleshy cotyledons and a small superior radicle, usually deformed from mutual pressure, the seeds being most frequently polyembryonous.

GARDEN PEA: *Pisum sativum.*

CLASS, Dicotyledons.   *Division*, Calycifloræ.

A weak climbing annual herb, with alternate stipulate compound leaves ending in tendrils, and irregular (papilionaceous) flowers.

ROOT fibrous, branching.

STEM weak, climbing, slightly branched, glabrous.

LEAVES cauline, alternate, pinnate (bi-tri-jugate), terminating in tendrils (metamorphosed leaflets) ; leaflets ovate, entire, glabrous, glaucous ; *stipules* foliaceous, ovate-cordate, slightly crenate.

STAMENS large, irregular (papilionaceous), in two- or three-flowered, axillary, pedunculate racemes.

CALYX gamosepalous, five-toothed, bilabiate, persistent.

COROLLA papilionaceous, white ; *vexillum* large, broadly obcordate, erect ; *alæ* roundish, converging, shorter than the compressed, curved *carina.*

STAMENS perigynous, decandrous, diadelphous ; *filaments* subulate above ; *anthers* two-celled, dehiscing longitudinally.

PISTIL apocarpous, monogynous ; *ovary* superior, oblong, compressed, one-celled ; *style* terminal, subfalcate ; *stigma* simple ; *ovules* few, attached to the ventral suture.

FRUIT a legume ; seeds few (3-9), globose, exalbuminous, with a coriaceous, glabrous testa.

TAMARIND : Tamarindus indica.

CLASS, Dicotyledons.   Division, Calycifloræ.

A large much-branched (introduced) tree, with broadly spreading crown, alternate multifoliolate leaves and yellowish or red-striped flowers in simple or panicled racemes.

EXTREMITIES distichously branched, often rugulose, glabrous, or at first thinly pubescent or puberulous.

LEAVES alternate, distichous, abruptly pinnate, glabrous ; *leaflets* usually in nine to twenty-one pairs, small (one-third to one-quarter inch long), coriaceous, oblong, obtuse, reticulate, subsessile ; *stipules* linear, early caducous.

FLOWERS small, in simple or panicled, terminal or lateral racemes ; *bracts* obovate-elliptical (one-quarter to one-third inch), concave, caducous ; *bracteoles* valvate, enclosing the early bud, pubescent, caducous ; *pedicels* spreading, equalling or shorter than the flowers.

CALYX gamosepalous ; *tube* narrowly funnel-shaped (*infund-ibuliform*) ; *limb* quadripartite, *segments* imbricate, entire, subequal in length, submembranous.

PETALS three (one posterior, two lateral), oblong or obovate-oblong, subequal, equalling or but slightly exceeding the calyx, narrowed to the base or shortly clawed ; two anterior petals minute subulate or squamiform.

STAMENS three, anterior, alternating with minute or obsolete staminodes ; *filaments* connate nearly half their length ; *anthers* oblong, versatile.

PISTIL apocarpous, monogynous ; *ovary* stipitate ; *style* rather short, equalling the stamens ; *stigma* terminal obtuse, slightly thickened ; *ovules* eight, ten, or more.

FRUIT an oblong or linear-oblong, subterete or slightly compressed, curved or nearly straight indehiscent legume ; outer layer of *pericarp* thinly crustaceous rugulose or nearly smooth, inner pulpy, fibrous, enveloping the seeds.

SEEDS roundish or obovate, compressed, with a thick shining testa, each side marked with a large faintly-defined areole ; *albumen* o.

GARDEN ZINNIA : *Zinnia elegans.*

CLASS, Dicotyledons.   *Division*, Corolliflorae.

An annual herb, with opposite entire leaves and terminal solitary heads of orange, scarlet, purple, rose, or white florets.

ROOT annual, slender, tapering, giving off numerous wiry, fibrous branches.

STEM erect, one and a half to four feet high, simple, or with one or two pairs of opposite ascending branches, terete, sparsely hirsute.

LEAVES cauline, opposite or subopposite, sessile, amplexicaul, ovate-oblong or ovate-elliptical, base cordate, apex acute or broadly pointed, entire, minutely setulose, scabrid or glabrescent, membranous, three- to five-nerved, the lateral nerves evanescent above ; exstipulate.

CAPITULA terminal solitary pedunculate heterogamous; *peduncles* erect, exceeding the upper leaves, appressed-tomentose or pubescent above; *involucre* hemispherical of numerous imbricate unequal obovate or obovate-oblong, very obtuse, scarious glabrous bracts margined with black above, the outer scales appressed, the inner at length with spreading or recurved apices; *receptacle* conical at length elongate, paleaceous; *paleæ* submembranous, lanceolate or linear conduplicate equalling the florets, with a minutely laciniate or dentate coloured apex; *florets* numerous, of the *disk* regular, hermaphrodite, of the *ray* irregular, pistillate.

CALYX gamosepalous, adherent; *limb* obsolete.

COROLLA of the *ray-florets* ligulate spreading obovate or obovate-oblong entire subpersistent, continuous below into the adnate calyx-tube; of the *disk-florets* tubular, five-toothed, tube abruptly dilated and articulated below, teeth spreading, shortly villous.

STAMENS pentandrous, epipetalous; *filaments* filiform; *anthers* syngenesious, linear, two-celled, dehiscing longitudinally, unappendaged below.

PISTIL syncarpous; *ovary* inferior, one-celled, uniovulate; *style* filiform; *stigma* bifid; *ovule* anatropous, erect.

FRUIT an achene; *pappus* o; achenes of the *ray* obovate much compressed shortly hispid more especially on the margins, continued above into the persistent base of the ligule; achenes of the *disk* obovate-oblong, compressed, entire or shortly bidentate above, thinly scabrid; *seed* solitary, erect, exalbuminous; *embryo* with an inferior radicle.

ROSE PERIWINKLE: *Vinca rosea.*

CLASS, Dicotyledons.   *Division,* Corollifloræ.

A perennial herb with opposite simple leaves, and axillary conspicuous rose or white flowers.

STEM usually woody and branching below, erect terete pubescent or puberulous.

LEAVES cauline, opposite, membranous, obovate or elliptic-oblong, obtuse mucronulate entire shortly pubescent or at length nearly glabrous above, narrowed below into the short *petiole.*

FLOWERS axillary in pairs, or sometimes solitary, subsessile or pedicels shorter than the calyx, regular, hermaphrodite.

CALYX inferior, pubescent, divided nearly to the base into five subulate segments many times shorter than the corolla-tube.

COROLLA hypogynous, hypocrateriform; *tube* cylindrical, constricted above and five-tubercled at the mouth; *limb* broadly spreading, of five obliquely-obovate or rotundate lobes, contorted in æstivation.

STAMENS epipetalous, pentandrous, alternate with the corolla-lobes; *anthers* subsessile, linear-oblong, subacute, obtusely sagittate at base, two-celled, polleniferous throughout.

PISTIL syncarpous, monogynous, dicarpellary, the carpels distinct in the *ovary*, united above in a single slender *style;* *stigma* slightly constricted horizontally, crowned by a hairy tuft; *ovules* indefinite, attached to the ventral suture of the carpels.

FRUIT of two narrow cylindrical follicles (half to one and a half inches long); *seeds* indefinite, oblong-cylindrical, minutely tubercled, unappendaged, albuminous; *embryo* cylindrical, radicle superior terete equalling or exceeding the cotyledons; *albumen* fleshy, confluent with the testa.

INDIAN WILLOW: *Salix tetrasperma.*

CLASS, Dicotyledons.   *Division,* Achlamydeæ.

A deciduous spreading tree, with alternate simple leaves and amentaceous diœcious flowers.

BRANCHES terete, glabrous, or the annual shoots silvery-pubescent.

LEAVES alternate petiolate membranous or thinly coriaceous, varying from ovate- or oblong-lanceolate to oval and linear-lanceolate, usually finely acuminate, rounded or more or less narrowed at the base, entire or minutely serrulate, glabrous above or at first appressed-sericeous, glabrous or thinly appressed-sericeous and opaque or glaucescent beneath; *stipules* ovate or oblong deciduous.

FLOWERS diœcious, achlamydeous, in axillary pedunculate elongate cylindrical occasionally rather lax silky or pubescent catkins; *peduncle* with or without a few reduced leaves : STAMINATE flowers with usually six to eight stamens in the axil of minute ovate bracts; *filaments* filiform, *anthers* minute, rotundate, two-celled ; *glands* two, anterior and posterior, minute, fleshy, inserted on the receptacle : PISTILLATE flowers in the axil of minute silky or tomentose bracts ; *ovary* shortly pedicellate, lanceolate-ovoid, pubescent or glabrous ; *glands* fleshy, embracing the pedicel ; *stigmas* two, subsessile ; *ovules* few, on two parietal placentas.

FRUIT a two-valved capsule, equalling or exceeding the pedicel ; *seeds* few (about four), comose, exalbuminous.

CULTIVATED RICE : *Oryza sativa.*

CLASS, Monocotyledons.   *Division*, Glumiferæ.

An annual cereal with an erect or slightly drooping narrow paniculate inflorescence.

ROOT fibrous.

STEM (*culm*) erect, or prostrate below and rooting at the nodes, jointed, terete, striate, glabrous.

LEAVES cauline, alternate, sheathing, linear finely acuminate, longitudinally-nerved, striate, more or less distinctly scabrous with minute setæ directly forwards ; *ligule* prominent, lanceolate, membranous or scarious.

INFLORESCENCE a panicle, narrow with short ascending lateral branches or more diffuse, branches wiry, angular or sulcate, hispid, or scabrous ; spikelets pedicellate one-flowered oblong compressed.

OUTER GLUMES two, minute, nearly equal, subulate, three to five times shorter than the spikelet.

FLOWERING GLUME navicular, coriaceous, laterally compressed, carinate, obscurely five-nerved, apiculate or aristate, the scabrous awn often many times longer than the spikelet, thinly setulose-hispid externally at least on the keel, or glabrate.

PALE equalling or nearly equalling the flowering glume, coriaceous, laterally compressed, obscurely three-nerved, mucronate or apiculate, thinly and minutely appressed-hispid or glabrate.

STAMENS hypogynous, hexandrous; *filaments* filiform; *anthers* versatile linear bilocular dehiscing longitudinally.

LODICULES two, glabrous fleshy.

PISTIL syncarpous; ovary globose or ovoid glabrous one-celled; *styles* two, slender; *stigmas* plumose; *ovule* solitary.

FRUIT a free linear or oblong caryopsis, closely invested by the persistent flowering glume and pale; *embryo* oblique, at the base of horny albumen.

N.B.—Rice occurs under many varieties, as do several of the plants here briefly described. The varieties of Rice differ in the presence or absence of an awn to the flowering glume, the colour of the glumes, the form of the caryopsis, and in other trivial characters.

II.

LIST OF SOME WORKS ON INDIAN BOTANY

*Several of these are rare or out of print, but may be occasionally
picked up second-hand.*

*There is no good work on the general botany of India. Several
named in the following enumeration are excellent so far as
they go, but most of them either apply to the botany of a
limited area, or are incomplete or out of date; others again
are hardly scientific.*

HOOKER and THOMSON.—"Flora Indica," vol. i. 8vo.; London,
1855. Descriptions of Genera and Species, from Ranunculaceæ
to Fumariaceæ, with a unique Essay prefixed on the Geographical relations of the Indian Flora. Under the title of "Præcursores ad Floram Indicam," the same authors continue, in
the "Journal of the Linnean Society of London," systematic
reviews of other Natural Orders : Campanulaceæ, Saxifrageæ
and allies, Crassulaceæ, Caprifoliaceæ, Balsamineæ, and Cruciferæ have been already published.

WIGHT and ARNOTT.—"Prodromus Floræ Peninsulæ Indiæ
Orientalis," vol. i. 8vo.; London, 1834. Descriptions of Genera
and Species of the Indian Peninsula, from Ranunculaceæ to
Dipsaceæ.

ROXBURGH.—"Flora Indica." Three vols. 8vo.; Serampore,
1832. The third volume, edited from posthumous manuscripts

only, occasionally contains the same species described twice over under different names.

WIGHT.—" Illustrations of Indian Botany ; " or Figures illustrative of each of the Natural Orders of Indian Plants described in the author's " Prodromus Floræ Peninsulæ Indiæ Orientalis," with Observations, &c.   Two vols. 4to. ; Madras, 1838–1841.

WIGHT.—" Icones Plantarum Indiæ Orientalis," or Figures of Indian Plants.   Six vols. 4to. ; Madras, 1838–1853.   A collection of upwards of 2,000 uncoloured lithographs, with brief descriptions.

THWAITES.—" Enumeratio Plantarum Zeylaniæ."   One vol. 8vo.; London, 1864.   An enumeration of the Flowering Plants and Ferns of Ceylon, with descriptions of many new species.

ROYLE.—" Illustrations of the Botany, &c. of the Himalaya." Two vols. folio ; London, 1839; with 100 coloured Plates.

ROXBURGH.—" Plants of the Coast of Coromandel."   Three vols. large folio ; London, 1795–1819 ; with 300 coloured Plates.

WALLICH.—" Plantæ Asiaticæ Rariores."   Three vols. folio ; London, 1830–1832 ; with 295 coloured Plates.

WALLICH.— " Tentamen Floræ Nepalensis."   One vol. folio; Calcutta, 1824–1826; with fifty Plates of select Nepal Plants.

HOOKER.—" Illustrations of Himalayan Plants."   One vol. large folio ; London, 1855 ; with twenty-five coloured Plates of remarkable plants of the Sikkim-Himalaya.

HOOKER.—" Rhododendrons of Sikkim-Himalaya."   One vol. folio ; London, 1849 ; with thirty Plates.

JACQUEMONT.—" Voyage dans l'Inde pendant les Années 1828–1832."   Botany by Cambessèdes and Decaisne.   One vol. folio ; Paris, 1841–1844 ; with 180 Plates.

DALZEL and GIBSON.—" Bombay Flora."   One vol. 8vo.; Bombay, 1861 (as to species only).

KLOTZSCH and GARCKE.—" Die botanischen Ergebnisse der Reise . . Prinzen Waldemar von Preussen."   Berlin, 1862. One vol. 4to. with 100 Plates, and Descriptions of Himalayan Plants.

DRURY.—" Handbook of the Indian Flora." Two vols. 8vo. ; Travancore, 1864–1866.

GRIFFITH.—" Posthumous Works ;" Calcutta, 1847–1854 ; including " Palms of British India," one vol. folio ; " Notulæ ad Plantas Asiaticas," being Miscellaneous Observations on Indian Plants, three vols. 8vo. ; " Journal of Travels in India," one vol. 8vo. ; and " Itinerary Notes," one vol. 8vo.

BEDDOME.—" Ferns of Southern India ;" Madras, 1863, one vol. 4to. ; with 271 Plates of Ferns (with descriptions) of the Madras Presidency : also " Ferns of British India ;" Madras, 1865. The latter work in progress ; exclusive of species figured in the previous work.

DON, D.—" Prodromus Floræ Nepalensis." One vol. 12mo. ; London, 1825.

MITTEN.—" Mosses of the East Indies." Journal of the Linnean Society, iii. 1859 (*Supplement*).

In the same Journal are numerous Monographs and Essays relating to Indian Botany, by Hooker and Thomson, Lindley, Edgeworth, Anderson, Aitcheson, Oliver, and others.

Many plants of North-Western India, of Affghanistan and Beloochistan, are described in Boissier's " Flora Orientalis," vol. i. 8vo., Basileæ, 1867 (*Thalamifloræ*); and " Diagnoses Plantarum Orientalium Novarum," two series, two vols. Geneva, 1842–1859. See also Hooker's " Journal of Botany" and " Kew Journal of Botany" for Miscellaneous Memoirs on Indian Botany, by Bentham, Stocks, and others ; Cleghorn's " Forests and Gardens of Southern India," one vol. 8vo. London, 1861 ; " Pharmacopœia of India," one vol. 8vo. London, 1868 ; Hooker's " Himalayan Journals," two vols. 8vo. London, 1854 ; Thomson's " Western Himalaya and Thibet," one vol. 8vo. London, 1852 ; Bentham's " Flora Hongkongensis," London, one vol. 8vo., 1861.

INDEX AND GLOSSARY.

A.

Abortion, imperfect or rudimentary development.
abrupt, applied to organs terminating suddenly.
Abrus, 197.
Absorption, 19.
Acacia, 199.
Acanthaceæ, 255.
Acanthus Family, 255.
acaulescent, apparently stemless, 68.
accrescent, applied to parts of the calyx or corolla which persist and enlarge after flowering, 188, 192-3.
Acer, 191.
Achene, 101.
Achimenes, 246.
Achlamydeæ, 60.
achlamydeous, without either calyx or corolla, 41.
Achras, 236.
acicular, 77.
Aconitum, 148.
Acorn, 98.
Acorus, 308.
Acotyledons, 342.
Acrotrema, 149.
Actæa, 148.
Aculei, prickles; *aculeate,* prickly.
Acumen, a long narrow point; *acuminate,* having an acumen.
acute, 79.
Adam's Needle, 73, 312.
Adansonia, 175.
Adhesion, 29.
Adiantum, 345.
adnate, adherent; also applied to anthers with the filament prolonged up the back of the anther.
Adventitious roots, 69.
Ægiceras, 261.
Æginetia, 253, 254.
Ægle, 184.
Aerial roots, 69.
Æschynanthus, 246.
Æschynomene, 197.
Æstivation, 88.
Agave, 73, 312, 327.
Alæ, wings (of papilionaceous corolla, 196); *alate,* winged.
Albizzia, 198.
Albumen, 45.
albuminous, 45, 105.
Alburnum, sap-wood (of ebony), 235.
Algæ, 358.
Alismaceæ, 331.
Allamanda, 239.
Allium, 312.
Almond, 203.
Aloe, 312.
Aloe, American, 327.
Alopecurus, 336.
Aloysia, 259.
Alpinia, 322, 325.
Alsodeia, 163.
Alsophila, 347.
alternate, 73.
Alysicarpus, 197.
Amaranth Family, 268.
Amaranthaceæ, 268.
Amaryllidaceæ, 326.
Amaryllis Family, 326.
American Aloe, 73, 312.
Amomum, 325, 326.
Ampelideæ, 188.
amplexicaul, 80,
Amygdalus, 203.
Anacardiaceæ, 192.
Anagallis, 263.
anatropous, applied to ovules when inverted, so that the micropyle adjoins the hilum, and the organic base of the nucleus (where it is united to the coat or coats of the ovule) is at the extremity remote from the hilum, 26.
Andrœcium, the stamens of a flower collectively.

Andropogon, 203, 337, 340.
-androus, in composition, applying to the stamens.
Anemone, 148.
Anethum, 219.
Angiosperms, plants having the ovules fertilised through the medium of a stigma, 295.
Anise, 222.
Anisum, 222.
Annatto Family, 164.
annual, producing seed and dying in the first season.
annual zones (of wood), 118.
Anona, 151.
Anonaceæ, 151.
anterior, same as inferior, when applied to the relation of the parts of a flower to the axis.
Anther, 8.
Antheridia, 349.
Antheridium, the male organ of Cryptogams, corresponding to the anther of Phænogams.
Antiaris, 275.
Antirrhinum, 253.
apetalous, without petals (or corolla).
Apex, 79.
apiculate, with a small abrupt point (*apiculus*).
Apium, 222.
apocarpous, 12.
Apocynaceæ, 238.
Apostasieæ, 321.
apothecium, the fructification of Lichens; usually applied to the open, shield-like lisks bearing the sporanges or thecæ.
Apple, 201, 203.
Apricot, 203.
Aquilaria, 287.
Arachis, 197.
arachnoid, like cobwebs.
Araliaceæ, 222.
Archegonium, the female organ of Cryptogams, corresponding to the ovule or embryo   c of Phænogams, 349.
Ardisia, 261.
Ardisia Family, 260.
Areca, 301.
Arenga, 301.
Argemone, 159.
Argyreia, 247.
Aril, or *arillus*, a coat growing partially or wholly over the testa of certain seeds, developed from the funicle or micropyle. An aril developed from the micropyle is sometimes distinguished as an *arillode*.
aristate, having an awn (*arista*).
Aristolochia, 281.
Aristolochiaceæ, 281.
Armeniaca, 203.
Aroideæ, 305.

Arrowroot, 325.
Arrowroot Family, 322.
Artabotrys, 152.
Artichoke, 230.
Artichoke, Jerusalem, 230.
Artocarpus, 272, 275.
Arum Family, 305.
Arundo, 338.
Ascending axis, 15.
Asci, 356.
Asclepiadaceæ, 239.
Asclepias Family, 239.
Ash of plants, 20.
Asparagus, 312.
Aspidium, 343, 346.
Asplenium, 345.
Assimilation, 21.
Atriplex, 267.
Atropa, 252.
Aubergine, 251.
Auricle, an ear-like appendage.
auriculate, with auricles.
Avena, 339.
Averrhoa, 182.
Awn, 56.
Axil (of leaf), 3.
Axile placentation, 96.
axillary, 70.
Axis, 15.
Azadirachta, 186, 187.
Azalea, 233.

B.

Bael, 184.
Bakas, 255.
Balm of Gilead, 186.
Balsam, 181, 182.
Bamboo, 338.
Bambusa, 338.
Banana, 324.
Banyan, 272.
Baobab, 175.
Bark, 118.
Barley, 338.
Base (of leaf), 79.
Basella, 267.
Basidia, 356.
Basil, 29, 38, 257.
Bassia, 236.
Batatas, 248.
Bead-tree, 186.
Bean, 196, 197.
Bean-caper Family, 180.
Beech, 285.
Beet, 267.
Begonia Family, 216.
Begoniaceæ, 216.
Bell-flower Family, 231.
Benincasa, 215.
Benzoin Family, 233.
Bertholletia, 210.
Beta, 267.
Betel, 301.

Betle, 293.
Bhang, 273.
bi-, two, in composition.
bidentate, with two teeth.
biennial, producing seed and dying in the second season.
bifid, 79.
bifoliolate, with two leaflets.
Bignonia Family, 244.
Bignoniaceæ, 244.
Bikh poison, 148.
bilabiate, 39.
bilocular, two-celled ; applied to anthers and ovaries.
Bindweed Family, 247.
bipartite, 80.
Birthwort Family, 281.
biternate, 79.
Bixa, 164.
Bixaceæ, 164.
blade, 4.
Blechnum, 345.
Blumea, 229.
Bœhmeria, 29, 40, 270.
Boerhaavia, 265.
Bokul, 235.
Bombax, 176.
Borage Family, 248.
Boraginaceæ, 248.
Borassus, 301, 302.
Boswellia, 185.
Botrychium, 348.
Bougainvillea, 266.
Bowstring Hemp, 312.
Bracteole, the small bract of an individual flower of an inflorescence.
Bracts, *bracteate*, 87.
Bramble, 200.
Brasenia, 155.
Brasil-nuts, 210.
Brassica, 31, 159, 160, 161
Bread-fruit, 275.
Brinjal, 251.
Bromelia, 304.
Broom-rapes, 253.
Broussonetia, 275.
Bruguiera, 211.
Bryophyllum, 217, 218.
Buchanania, 192.
Buckthorn, 206.
Buckthorn Family, 205.
Buckwheat, 101, 269.
Buckwheat Family, 269.
Bud, 6.
Bulb, 71.
Bull-rush Family, 304.
Bunt, 356.
Burmannia, 322.
Burmanniaceæ, 322.
Buro-kanoor, 29, 52.
Burseraceæ, 185.
Butea, 195.
Butterwort Family, 254.

C.

Cabbage, 160.
Cactaceæ, 215.
Cadaba, 162.
caducous, applied to organs of the flower which fall off at or before the time of expansion, 31.
Cæsalpinieæ, 196.
Cajanus, 199.
Calabash-tree, 245.
Calamander, 235.
Calamus, 301.
Calceolaria, 254.
calceolate, slipper-shaped.
Calophyllum, 168.
Calosanthes, 244.
Calotropis, 240.
Calumba-root, 154.
Calyciflorae, 60, 194.
Calyptra, 354.
Calyx, 7.
Cambium, 116.
Camellia, 170.
Campanula, 231.
Campanulaceæ, 231.
campanulate, bell-shaped.
Camphor, 292.
campylotropous, applied to ovules when the nucleus and its coat are curved so as to bring the micropyle near to the hilum and to the organic base of the nucleus.
canescent, hoary with minute hairs, giving the surface a whitish hue.
Canna, 324.
Cannabis, 273.
Canthium, 227.
Cape Gooseberry, 100, 251-2.
Caper Family, 161.
Capitula, 36.
Capparidaceæ, 161.
Capparis, 162.
Capsule, 101.
Caraways, 222.
Carbon of plants, 21.
Carbonic acid, 21, 120.
Cardamoms, 326.
Cardiospermum, 191.
Carex, 334.
Carina, a keel (of papilionaceous corolla), 196.
carinate, keeled.
Carissa, 239.
Carlemannia, 226.
Carpels, 11.
Carpophore, 220.
Carrot, 219.
Carthamus, 230.
Carum, 222.
Caryophyllaceæ, 166.
Caryophyllus, 209.
Caryopsis, 339.

Caryota, 301, 302.
Cassava, 280.
Cassia, 196, 197, 198, 292.
Cassytha, 292.
Castor-oil Family, 276.
Castor-oil seed, 45.
Catechu, 199.
Catkin (or *Ament*), a deciduous spike, 41, 283.
Caudicle, 317.
cauline, 4.
Cedar, 297.
Cedar, W. Indian, 187.
Cedrela, 187.
Cedrus, 297.
Celastraceæ, 204.
Celastrus, 204.
Celery, 222.
Cell-contents, 111.
Cells, 108.
Cellular plants, 354.
Cellular system, 114.
Cellulose, 112.
Celosia, 268.
centrifugal, same as definite, applied to an inflorescence.
centripetal, same as indefinite, applied to an inflorescence.
Cephaëlis, 226.
Ceratopteris, 347.
Cereals, 339.
cernuous, pendulous, overhanging.
Cetraria, 357.
Chalaza, the part of an ovule where the base of the nucleus is united to its coat.
Chamærops, 301.
Chaste-tree, 259.
Chavica, 293.
Cheer Pine, 294.
Cheersullah, 294.
Chenopodiaceæ, 266.
Chenopodium, 266.
Cherry, 203.
Chestnut, 284.
Chlorophyll, 113.
Chloroxylon, 7
Chocol , 176.
Chrysophyllum, 236.
Churrus, 273.
Cicca, 280.
Cicer, 196.
Cinchona, 226.
Cinchoraceæ, 230.
Cinenchyma, branching or simple vessels containing white or coloured fluid (milk-sap).
Cinnamomum, 291.
Cinnamon, 291, 292.
circinate, 343.
circumscissile, dehiscing transversely; applied to capsular fruits.
Cissampelos, 154.

Citron, 6, 184.
Citrullus, 214, 215.
Citrus, 5, 182.
Citrus aurantium, described, 365.
Cladonia, 357.
Class, 45.
Classification, 123.
Clausena, 184.
Claw of petal, 89.
Clearing nuts, 242.
Clematis, 147, 148.
Clerodendron, 260.
Clove, 209.
Clove Pepper, 209.
Clubmoss Family, 349.
Coat of ovule, 26.
Coccinia, 215.
Cocculus, 154.
Coccus, the one-seeded carpel of a syncarpous fruit, the carpels of which separate from each other when ripe.
Cockscomb, 268.
Cocoa-nut, 100, 300.
Cocoa-tree, 176.
Cocos, 300, 301.
Coffea, 226.
Cohesion, 29.
Coix, 336.
Collective fruits, 102.
Colocasia, 29, 47, 305, 307.
Colocynth, 215.
Columella, the central axis in the sporange of Mosses.
Column, 317; of Orchids, 51.
Coma, 241.
Combretaceæ, 203.
Combretum Family, 203.
Commelyna, 313.
Commelynaceæ, 313.
Commissure, 221.
complete, applied to flowers when calyx, corolla, stamens, and pistil are present.
Compositæ, 227.
Composite Family, 227.
Compound leaves, 75, 78.
Conceptacle, a closed cavity containing fructification in Cryptogams; of Fucus, 358.
conduplicate, folded down the middle.
Cone, 102.
Confervoideæ, 358.
Coniferæ, 294.
Conium, 221.
Connaraceæ, 193.
Connarus Family, 193.
connate, 80.
Connective, the portion of the anther connecting the pollen-bearing lobes.
Constituents of plants, 20.
contorted, twisted.
Convolvulaceæ, 247.
Convolvulus, 247, 248.
Copal, Indian, 172.

D D

LONDON :

R. CLAY, SONS, AND TAYLOR, PRINTERS.

BREAD STREET HILL.